Computer Applications in Manufacturing

Computer Applications in Manufacturing

Thomas G. Gunn

Industrial Press Inc.
200 Madison Avenue / New York, New York 10157

Library of Congress Cataloging in Publication Data

Gunn, Thomas G.
Computer applications in manufacturing.

Bibliography: p.
Includes index.
1. Production management—Data processing.
2. Production control—Data processing. I. Title.
TS155.G924 658.5'0028'54 81-6544
ISBN 0-8311-1087-2 AACR2

First printing

Acknowledgments

I would like to acknowlege the generous support of Arthur D. Little, Inc., and in particular that provided by Thomas D. Rowen during my preparation of this book.

In addition, Sandra Hickey provided me with much encouragement and help in the final and critical stages of the manuscript preparation.

My thanks to these two people particularly, and to all my other friends who gave me encouragement to write this book.

Preface

In the last decade a quiet revolution has been occurring in manufacturing—the application of computers to manufacturing. This revolution has been made possible by the increasingly lower cost and faster speed of computing and information storage and retrieval. It has been driven by rising costs of everything used in manufacturing—land, labor, capital, and materials. A further driving force for the adoption of such systems has been management's need for greater *control* over its business in order to enhance productivity and increase profits.

There are three main aspects of this revolution in manufacturing and manufacturing management:

> 1 / The first of these is the tremendous growth of the strategic planning process. Turbulent business conditions, accompanied by an increase in the number of MBA's who have been taught strategic planning, have helped make Strategic Planning an accepted legitimate business function. While Strategic Planning was originally created and used primarily by marketing and financial people, top corporate management is now realizing that Strategic Planning in Manufacturing—Manufacturing Resource Planning—is just as valuable as Strategic Planning in Marketing or Finance. Therefore, Manufacturing Strategy has become an integral part of the company's total planning picture.
>
> 2 / The rapid growth of CAD (computer-aided design) and CAM (computer-aided manufacturing). CAD is involved in the product design and engineering phase of manufacturing. CAM—the outgrowth of the marriage of NC (numerical control) machinery to on-line computers—is concerned with the application of computers to many different aspects of the manufacturing or part machining process.
>
> 3 / Concurrent with the above, the explosive growth and utilization of modern manufacturing planning and control knowledge. The American Production and Inventory Control Society (APICS) has been the main proponent of this development. It is interesting that APICS members have led the way with new ideas on Manufacturing Planning and Control Systems—not graduate business school professors, or management services division of "Big Eight" accounting firms, or the major consulting firms. The revolution has come from the *users!* Indeed, APICS was just formed in 1957, and today—midway through 1981—has over 35,000 members.

Most of the developments regarding APICS-related information on Manufacturing Planning and Control Systems, the advances that include Manufacturing in the Strategic Planning Process, and developments in Computer-Aided Design and Manufacturing are very recent. To my knowledge, no one book has yet integrated these subjects in a "state of the art" textbook. I hope this book will serve that function. It is written primarily as an MBA level or an advanced undergraduate level text. Yet, I have also tried to write it so that it can be read and understood by the executive management and functional managers of manufacturing companies.

I am treating the computer's role in these Manufacturing Planning and Control Systems as that of an input/output data processing box. While I've included a chapter on Data Processing (DP) and Computers, I'll leave the technical aspect of DP to those much more qualified. From a data processing viewpoint, I am assuming throughout this book that the data-base concept of data storage and handling is here to stay. Therefore I will avoid talking about files as separate entities, i.e., the bill of materials file or routing file, and instead discuss the data associated with functional entities such as bills of materials or routings as a subset of the company's overall data base.

I have had to establish one further convention while writing this book since MRP now has several possible meanings. Big MRP is commonly known as Manufacturing Resource Planning. Little MRP was the first MRP—Material Requirements Planning. Yes, you guessed it. Now there is MRP II (!) where the company's accounting and financial systems are tied in with the company's MRP system. In this book, I will refer to Manufacturing Resource Planning as MRP, and use mrp for Material Requirements Planning, and leave MRP II as is.

I would welcome suggestions from readers or professors using this book as a text as to any improvements in format, presentation sequence or error corrections, that I could make to the second edition.

It has never been more important for executives and new students of manufacturing to understand the benefits of computer applications in manufacturing. Not only can the computer enable many management or execution functions to be accomplished more efficiently, i.e., with greater *productivity,* but the use of computers in manufacturing also allows more effective management planning, execution, and control.

There is growing evidence that the extent of computer applications in a manufacturing company is a major determinant of a company's success in terms of financial strength and market dominance. For this reason, an understanding of computer-integrated manufacturing is vital for today's businessman.

Computer Applications in Manufacturing is designed to integrate many aspects of computer applications into a modern, comprehensive, and practical text covering Manufacturing Planning and Control Systems, CAD and CAM.

Contents

1

A Manufacturing Overview

According to the U.S. Department of Labor's Bureau of Labor Statistics, there are currently 350,000 manufacturing firms in the United States. These manufacturing firms employ 21.5 million people, or approximately 21% of the total U.S. labor force. Manufacturing as an industry has the highest level of value added of all U.S. industrial sectors—more than the wholesale and retail trades, the transportation sector, or the real estate industries.

Manufacturing

A brief review of the subject of manufacturing and the typical manufacturing company follows, but, first, a definition of the word "manufacturing" is needed. For the purposes of this book, manufacturing is defined as the making of a product suitable for use as a component or end product—where raw or semifinished materials are processed through a series of operations in which value is added and, ultimately, a more-complete and useful product emerges. The manufacturing company not only must manufacture the product, but, in a broader sense, it must also be responsible for designing the product; planning, controlling, and engineering the manufacturing process; and moving the produced goods to the customer.

Manufacturing companies can have a variety of different organizational structures. Regardless of the actual arrangement of reporting relationships, the general *functions* that must be covered—whether explicitly or implicitly—by the organizational structure in a typical manufacturing company are:

Research and development
Design engineering
Manufacturing engineering
Marketing and market research
Sales and advertising
Personnel/human resources
Legal
Public relations
Finance
Industrial engineering
Quality control
Strategic planning
Production operations
(includes materials control and production control)
Data processing
Distribution
Purchasing
Accounting

Discrete-Part Versus Process Manufacturing

Manufacturing can be separated into two general areas— process and discrete-part industries—depending on the products produced. Process industries add value to materials by changing their properties. These industries are usually concerned with chemical reactions and the physical extraction, separation, or mixing of ores, liquids, and other materials. Discrete-product industries produce parts from raw materials. Later, these parts are usually combined into products designed to serve a distinct functional purpose.

In both discrete and process industries, parts can be made continuously, or in specific quantities known as batches or lots. In this book, we will be concerned primarily with discrete-product manufacturing.

A Manufacturing Company's Financial Picture

It is worthwhile to examine a typical manufacturing company's financial statements. Examples of a sample balance sheet and income statement are shown in Figs. 1-1 and 1-2. Note on the balance sheet the proportion of total assets that is inventory whether it be work in process, raw material, or finished goods. It is not uncommon for this inventory to be 30–50% of current assets for a representative manufacturing company.

Balance Sheet

Par Manufacturing Corporation
December 31, 1980
(in Thousands of Dollars)

Assets	
Current Assets	
Cash	$ 50
Marketable securities	180
Receivables	735
Inventories	507
Prepayments	2
Total Current Assets	1474
Investments	
Investments in jointly owned companies	148
Investments in subsidiaries	123
Other	67
Total Investments	338
Property	
Land, buildings, machinery, and equipment	$1985
Less: Accumulated depreciation and depletion	(921)
Net Property	1064
Other Assets	
Long-term receivables	81
Goodwill	79
Other	46
Total Other Assets	206
Total Assets	3082

Fig. 1-1. A typical manufacturing company's balance sheet.

Liabilities and Stockholder's Equity

Current Liabilities	
Notes payable	109
Current maturities of long-term debt	12
Accounts payable and accrued liabilities	543
Customer advances	60
Income taxes	115
Total Current Liabilities	839
Other Liabilities	
Long-term debt	376
Deferred income taxes	199
Other	71
Total Other Liabilities	646
Stockholders's Equity	
Cumulative preferred stock	2
Common stock	35
Additional paid in capital	580
Retained earnings	985
Less common stock in treasury	(5)
Total Stockholders's Equity	1597
Total Liabilities and Stockholders's Equity	3082

Fig. 1-1 *(continued)*

An improved manufacturing planning and control system allows better purchasing and inventory control. In turn, a one-time reduction in inventory level can be realized as a benefit of these improvements. This reduction in inventory level represents a cost avoidance or savings, since in a given period of time, we can produce the same products by turning the inventory more often.

If this one-time inventory cost avoidance could be added to the bottom

Income Statement

Par Manufacturing Corporation
for year ending December 31, 1980
(in thousands of dollars)

Revenue	
Net sales	$3380
Costs and Expenses	
Cost of goods sold	2535
Research and development	169
Selling, general and administrative	338
Interest and debt expense	17
	3059
Earnings before Provision for Taxes	321
Provision for Federal and State Taxes	136
Net Income	185

Fig. 1-2. A typical manufacturing company's income statement.

line of the income statement, its dramatic leverage on profit would become obvious. For example, assume a 10% inventory level reduction on a $20 inventory needed to support sales of $100 per year. Thus the year's cost avoidance is $2. If the "before" income statement was as shown in Fig. 1-3, to the income of $10 we then add the $2 inventory

Sales	$100
Cost of goods sold	80 (80%)
Net income before tax	20
Taxes	10 (50%)
Net income	10

Fig. 1-3. Example income statement *before* inventory reduction.

reduction cost avoidance since it represents an after-tax savings of money not invested in inventory. This results in a "net income" of $12. Now, keep the same cost and tax ratios and work back up the income statement to see what level of sales would be necessary to achieve the same "net income" of $12 (Fig. 1-4). Thus, we see that a 10% reduction in inventory level is the after-tax dollar "equivalent" of a 20% gain in sales in this particular example! (See Fig. 1-5.)

Sales	$120
Cost of goods sold	96 (80%)
Net income before taxes	24
Taxes	12 (50%)
Net income	12

Fig. 1-4. Example income statement *after* inventory reduction.

	Before	*After*
Sales	$100	$120
Cost of goods sold	80	96
Net income before taxes	20	24
Taxes	10	12
Net income	10	12

Fig. 1-5. Example income statement comparison.

While buried in conventional accounting procedures, this inventory cost avoidance savings is real and, of course, is reflected in the firm's increased inventory turns. Note that additional savings will also accrue from a reduction in inventory carrying cost charges. What executive today can ignore this direct impact on a company's profits that better manufacturing planning and control systems can offer?

Product Cost

Now, examine the product cost make up of a typical discrete product manufactured in the United States (Fig. 1-6). Note the dominance of the cost of materials in the total product cost, and the ratio of material to labor cost (normally at least 2 to 1, often 3 to 1). From this we can conclude that obtaining control over material cost offers the greatest leverage to improving the firm's total cost picture.

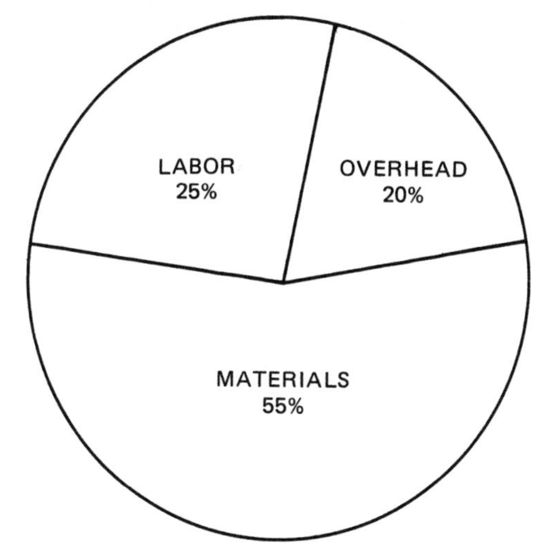

Fig. 1-6. Average manufactured-product cost breakdown.

Where Does Manufacturing Go Now?

There are two aspects to consider in plotting general developments that will occur in manufacturing in the next few decades. On the one hand, there is the planning and control aspect of manufacturing. Here, the concern is with the processing of information and the control of manufacturing operations according to manufacturing management's strategic plan.

In the past 10 years, the practice of strategic (long-range) planning has developed to the point that most companies recognize the value of such efforts, and either have a strategic planning system in place or are implementing one.

Computer-based financial and accounting system packages and marketing information systems are also beginning to be commonplace in manufacturing companies—the one unconquered field where enormous "bottom line" gains in productivity and in lowering product cost can be made is the manufacturing function, through the use of computer-based manufacturing planning and control systems.

On the other hand, manufacturing has an operations or process side where recent advances in computer-based applications are contributing significant gains in productivity and lower product cost. Computer-aided design (CAD) and computer-aided manufacturing (CAM) are fast becoming necessities for any U.S. manufacturer who wants to remain competitive in today's world markets.

Some Typical Rewards of Computer Applications in Manufacturing

Before going into the details of computer applications in manufacturing, some typical benefits that manufacturers have achieved in this area will be examined.

Pioneers in modern, computer-based manufacturing planning and control systems—e.g., Twin Disc, Black & Decker, and Steelcase—are now reaping the benefits of their first-class manufacturing planning and control systems. Typical benefits that have accrued to such users, measured from the "before manufacturing planning and control system" state, are shown on Fig. 1-7.

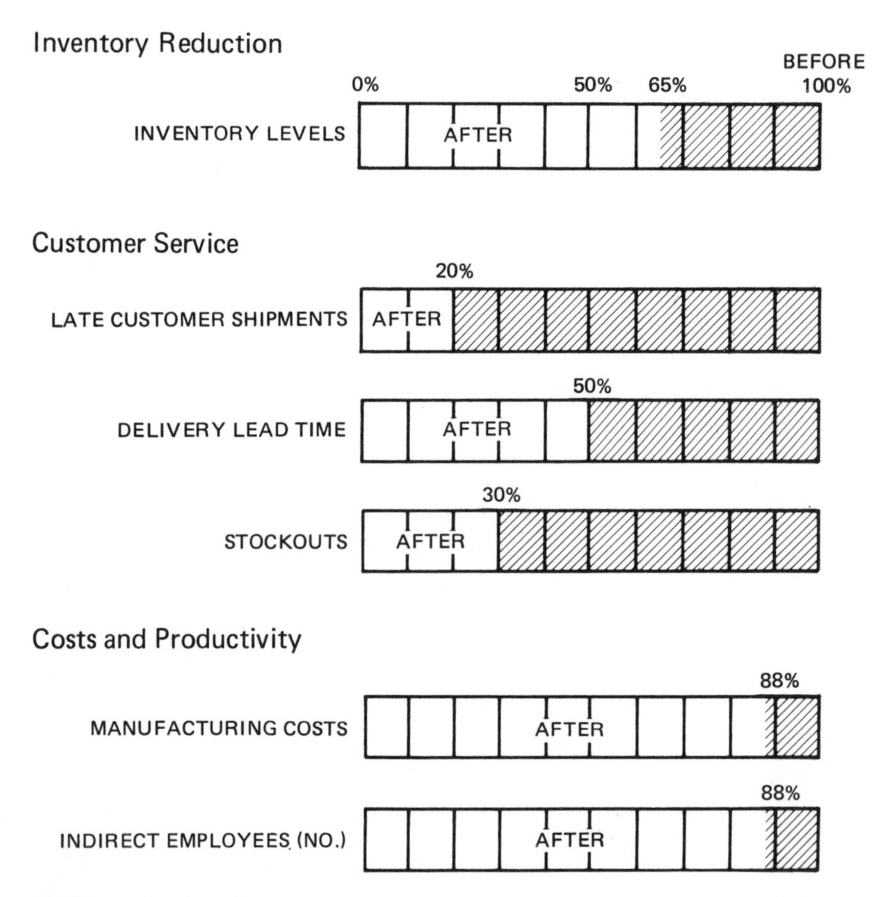

Fig 1-7. Typical benefits to users of computer-based manufacturing planning and control systems.

Similar productivity gains in CAD/CAM have also been achieved. In the CAD area, gains in productivity of 3 or 4 to 1 are not uncommon. In the CAM area, gains in productivity of 3 to 1 are the norm, in addition to other benefits such as greatly improved product quality.

With these kinds of payoffs in sight, and a base frame of reference from this chapter, we will explore computer applications in manufacturing.

2

The Bill of Materials

Every manufactured item has associated with it a bill of materials. This bill of materials, or product structure as it is sometimes called, is a list or description of every part that makes up the finished product. Without this list, it would be difficult for the people in production, purchasing, or materials management to know what parts and materials they have to buy and bring to the assembly line so that the finished product could be assembled.

Bill of Materials's Structure

There are several ways to refer to the bill of materials's structure, i.e., how the bill of materials is created and depicted. One way to describe bill of materials relationships is that there is an assembly that could be a finished good ("parent" or end item) made up of several parts that could be components or subassemblies ("children" or raw materials) (Fig. 2-1). Note that the product structure concept is built upon a level-by-level hierarchy where a higher-level part is the parent of one or more lower-level children as shown in Fig. 2-2.

From product structure diagrams such as these a listing of each part in the product, or a bill of materials, can be created. A bill of materials for a small hypothetical part is shown in Fig. 2-3.

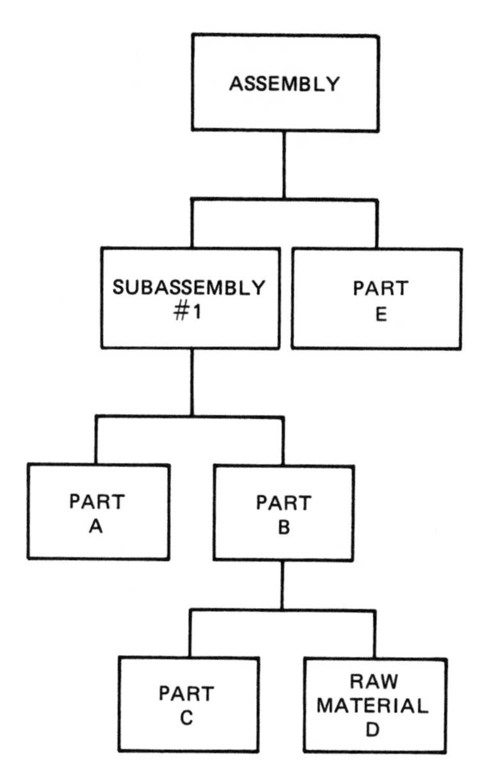

Fig. 2-1a. Product structure relationships.

This bill of materials informs Manufacturing, Purchasing, Finance, Engineering, and the customer (if it was included, for instance, in an instruction or parts manual) that Mounting Kit #47342 contains exactly the items and quantities shown.

Bill of Materials's Functions

The bill of materials results from the engineering part design process, and provides a way for the material needed per part, e.g., for Mounting Kit #47342, to be calculated by Purchasing or Manufacturing. It also allows Finance to cost out the material content of the kit, and price the material content as a part of the selling price calculation.

The bill of materials shows the parts that Manufacturing should assemble or put in a bag for shipment as Mounting Kit #47342; it also can be used to tell Manufacturing how the product is to be put together. Finally, it can tell the customer what parts should be found when a box containing Mounting Kit #47342 is opened.

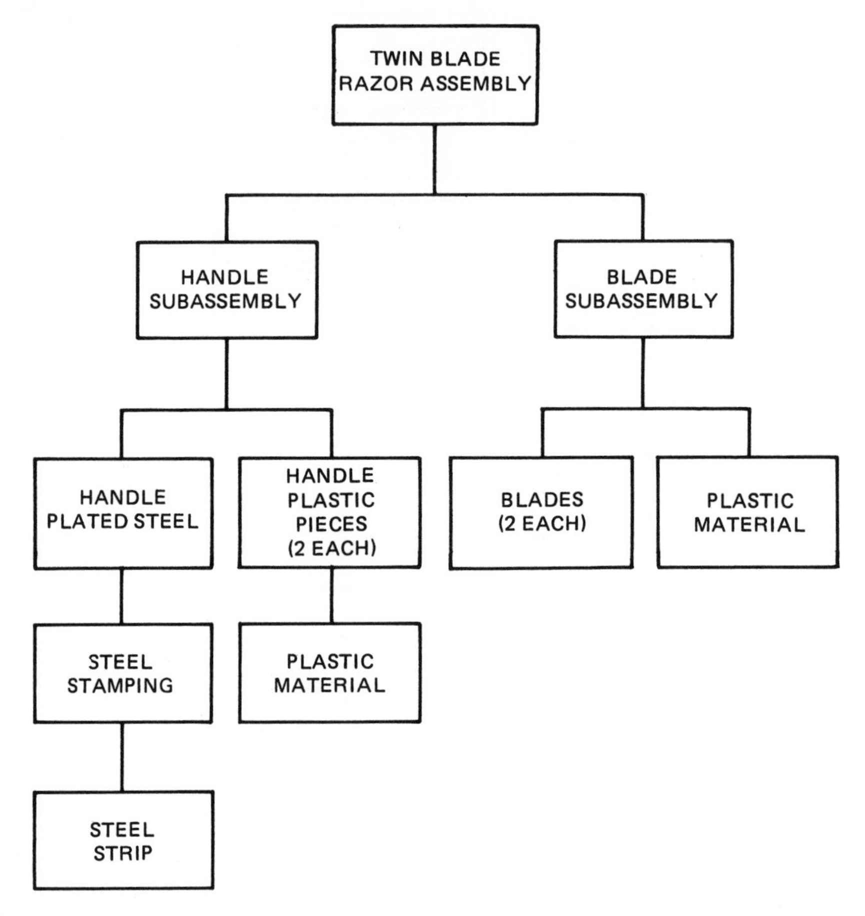

Fig. 2-1b. Product structure relationships.

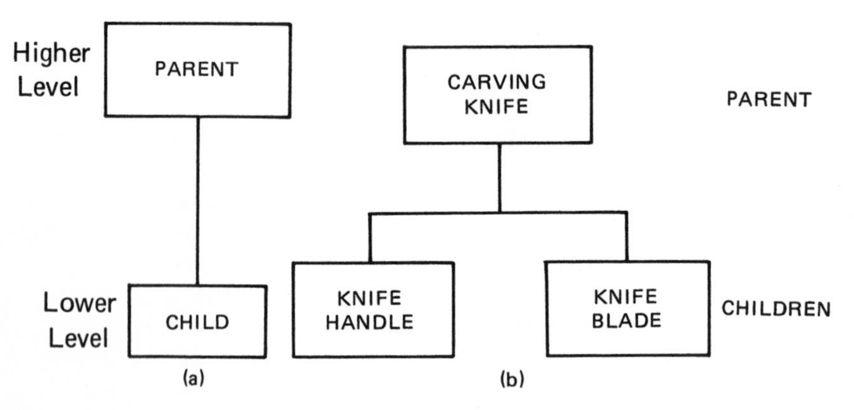

Fig. 2-2. The parent–child product structure relationship.

Single-Level Versus Multilevel Bill of Materials

The bill of materials shown in Fig. 2-3 is called a single-level bill of materials, which is pictorially represented in a product structure diagram in Fig. 2-4.

Bill of Materials

Part #47342 *Mounting Kit*

Item #	Description	Quantity
76504	Bracket, cast	1
64333	Bolt, $\frac{5''}{16} \times 24 \times 1''$	2
30751	Nut, $\frac{5''}{16} \times 24$	2
22479	Washer, flat $\frac{5''}{16}$	2
22842	Washer, lock $\frac{5''}{16}$	2

Fig. 2-3. Bill of materials for part #47342.

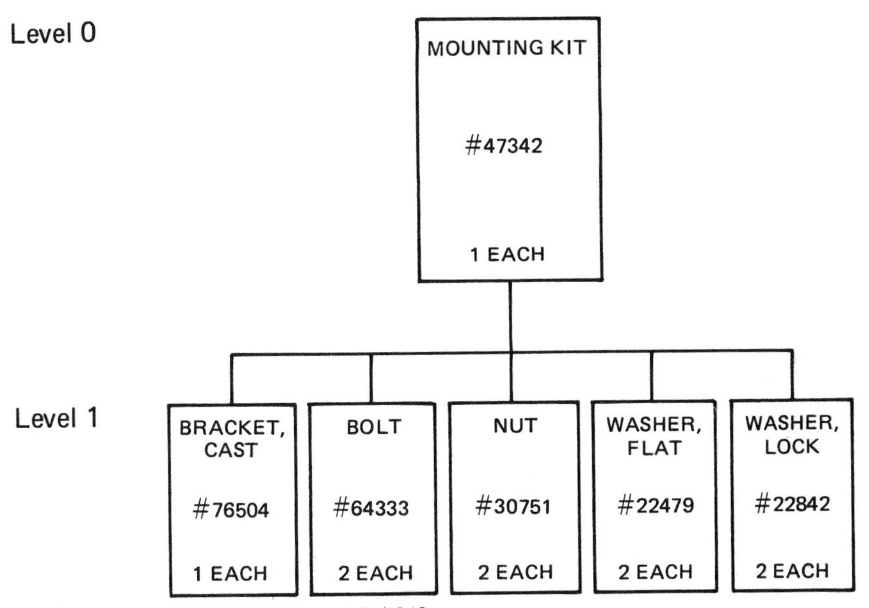

Fig. 2-4. Product structure for part #47342.

Note in Fig. 2-3 that there are no indentations of the five part numbers in the listing that makes up the kit. Now, contrast this with an indented bill of materials. For instance, if we change the mounting kit a little, the bill of materials would look like the one in Fig. 2-5. This more involved bill of materials is shown in a product structure diagram in Fig. 2-6.

Part #79111 Mounting Kit, Deluxe

Item #	Description	Quantity
47342	Mounting Kit, Basic	1
76504	Bracket, cast	1
64333	Bolt, $\frac{5''}{16} \times 24 \times 1''$	2
30751	Nut, $\frac{5''}{16} \times 24$	2
22479	Washer, flat, $\frac{5''}{16}$	2
22842	Washer, lock, $\frac{5''}{16}$	2
16935	Clamp assembly	1
88327	U bolt	1
30750	Nut, self-lock, $\frac{5''}{16} \times 24$	2

Fig. 2-5. Indented bill of materials for part #79111.

In American bill of materials structure convention, Level 0 is always the finished product or end item. As you proceed down in the bill of materials or product structure diagram, you arrive at lower levels but higher numbers. A Level 6 part is at a lower level than a Level 2 part. Bills of materials usually average four to eight levels deep in most industries.

Lower-level items may be used in many different end items or higher-level subassemblies. As an example, a particular bolt could be used in several subassemblies of an end item. However, it could also be used to attach a subassembly to the end item. Therefore, its lowest level could be as a bolt, independent of any subassembly.

Bill of materials levels can be shown either by implicitly indicating levels by indenting those parts that go into higher levels, as in Fig. 2-7, or by explicitly labeling levels, as in Fig. 2-8.

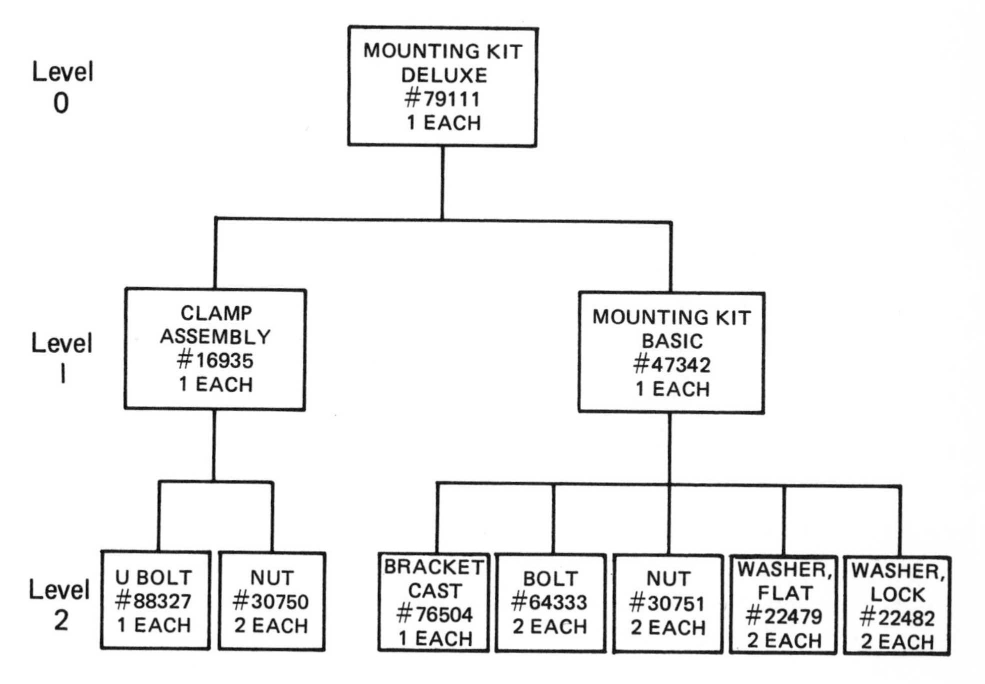

Fig. 2-6. Product structure for part #79111.

Item #	Description	Quantity
79111	Mounting Kit, Deluxe	1
47342	Mounting Kit, Basic	1
76504	Bracket, cast	1
64333	Bolt, $\frac{5''}{16} \times 24 \times 1''$	2
30751	Nut, $\frac{5''}{16} \times 24$	2
22479	Washer, flat, $\frac{5''}{16}$	2
22842	Washer, lock, $\frac{5''}{16}$	2
16935	Clamp assembly	1
88327	U bolt	1
30750	Nut, self-lock, $\frac{5''}{16} \times 24$	2

Fig. 2-7. Indented bill of material for part #79111.

Level	Part #	Description	Quantity
0	79111	Mounting Kit, Deluxe	1
1	47342	Mounting Kit, Basic	1
2	76504	Bracket, cast	1
2	64333	Bolt, $\frac{5}{16}''$ × 24 × 1″	2
2	30751	Nut, $\frac{5}{16}''$ × 24	2
2	22479	Washer, flat, $\frac{5}{16}''$	2
2	22482	Washer, lock, $\frac{5}{16}''$	2
1	16935	Clamp assembly	1
2	88327	U bolt	1
2	30750	Nut, self-lock, $\frac{5}{16}''$ × 24	2

Fig. 2-8. Bill of materials for part #79111 with the levels indicated.

For planning and computer-data-storage purposes, a summarized bill of materials allows a shorter listing as it combines requirements for identical parts in one line. The disadvantage of this single-level listing is that it does not show important structural relationships between the bill of materials parts as the indented bill of materials does. An example that illustrates this is shown in Figs. 2-9 and 2-10.

Part #3453 *Mounting Kit*

Item #	Description	Quantity
7786	Clamp Assembly	1
6002	U bolt, $\frac{5}{16}''$ ×24	1
2374	Nut, self-lock, $\frac{5}{16}''$ ×24	2
1015	Washer, flat, $\frac{5}{16}''$	2
5400	Bracket assembly	1
3374	Bolt, $\frac{5}{16}''$ ×24×2″	2
2374	Nut, self-lock, $\frac{5}{16}''$ ×24	2
1015	Washer, flat, $\frac{5}{16}''$	2

Fig. 2-9. Indented bill of materials for part #3453.

Part #3453 *Mounting Kit*

Item #	Description	Quantity
7786	Clamp Assembly	1
5400	Bracket Assembly	1
6002	U bolt, $\frac{5''}{16}$ × 24	1
3374	Bolt, $\frac{5''}{16}$ × 24 × 2"	2
2374	Nut, self-lock, $\frac{5''}{16}$ × 24	4
1015	Washer, flat, $\frac{5''}{16}$	4

Fig. 2-10. Summarized bill of materials for part #3453.

Note that the summarized bill of materials takes only six lines to list the parts contained in the end item; however, it shows no structural information. Thus, there is no way to tell which parts go together to form which subassemblies, or, indeed, that there are any subassemblies!

Bill of Materials Explosions

Bill of materials or part explosion is a term that means making a pass down through the bill of materials (from the highest level to the lowest level) in order to "explode out" gross part requirements. For example,

Level code	Part #	Quantity
0	A	
1	B	1
2	C	1
2	D	1
3	E	1
3	F	1
1	G	6
1	F	2
1	J	3

Fig. 2-11. Indented bill of materials for part A.

assume 10 part A's are to be made based on the bill of materials for part A that is given in Fig. 2-11. An *explosion* of this bill of materials would show that to produce 10 A's we need at level 1:

10 B's, 60 G's, 20 F's, and 30 J's

If no B or D subassemblies existed, we would need:

60 G's, 30 F's, 30 J's, 10 E's, and 10 C's.

Where-Used Reports and Bill of Materials Implosions

Utilizing data from all the company products's bills of materials, a complete listing of every place any one part is used in the company's entire product line can be obtained.

In a bill of materials parts *implosion,* a pass is made up the bill of materials to find out which parts "go into" what other parts or subassemblies. This parts *implosion* starts from the lowest level a part is utilized and proceeds *up* through every subassembly level of all products's bills of materials to all the end items (Level 0) where the part is utilized. The report generated in this process is generally called a *where-used report.* This can be very handy tool for determining the impact of design changes and as a way to implement part design commonizing. A sample where-used report is shown in Fig. 2-12.

Part #64731 *Retaining Pin*

Used in part #	Level
39739	3
75564	3
90028	2
11254	1
26621	0

Fig. 2-12. Where-used report for part #64731.

Engineering Changes

Manufactured parts are liable to go through several engineering or design changes, where new or modified parts must be phased in and out of a product's design. In addition, *new* products are constantly being created that must be phased into the manufacturing materials flow.

Typical circumstances where there can be changes to product parts are

New part design, e.g., changes to part shape or size

New part material or finish specification

New manufacturing process, e.g., heat treat, or forge instead of cast

New method of assembly, e.g., rivet instead of bolt

There are two fundamental methods of recording the timing of the engineering change in the bill of materials: the first is by manufacturing date; the second is by part serial number.

Engineering change timing involves the question of when the new part design is to be started in production. Is the design changed

1 / Immediately—from today on—to all newly manufactured products?

2 / When the old part runs out?

3 / Concurrent with the date of a new model release?

4 / Retroactively, to all models produced since last year, for instance?

Further design changes may be required for product liability reasons. In this case, it may be necessary to go back in the product's history to supply retrofit-part kits for old models or to supply replacement service parts for units already in the field. The answers to these questions have a considerable effect on what quantities of new or old parts will be needed and when.

In any case, a typical bill of materials showing an engineering change is shown in Fig. 2-13. Note that *two* dates or serial number fields are needed: one shows when the use of a part was *stopped,* the other shows when a new part number was *started.* The mrp logic would use the cast bracket in its requirements planning through 04-13-79, then automatically switch to using the forged bracket from 04-14-79 on.

#47342 Mounting Kit		*Effectivity Dates*	
	Quantity	Start	Stop
#76504 Bracket-cast	1	02-26-75	04-13-79
#76505 Bracket-forged	1	04-14-79	12-31-99
#64333 Bolt	2	02-03-67	12-31-99
#30751 Nut	2	11-17-66	12-31-99
#22479 Washer	2	09-15-77	12-31-99
#22842 Washer	2	04-09-74	12-31-99

Fig. 2-13. Bill of materials with effectivity dates for part #47342.

Phantom Bills of Materials

Frequently a product is sold with many options or in many similar models. When this occurs, instead of having, for example, 10 full-length bills of materials for each "standard option," it is often convenient to have a phantom bill of materials for the *basic* product, that is, for parts common to all 10 options. The term "phantom" is used because the part is never really sold as shown on the phantom bill of materials. Each of

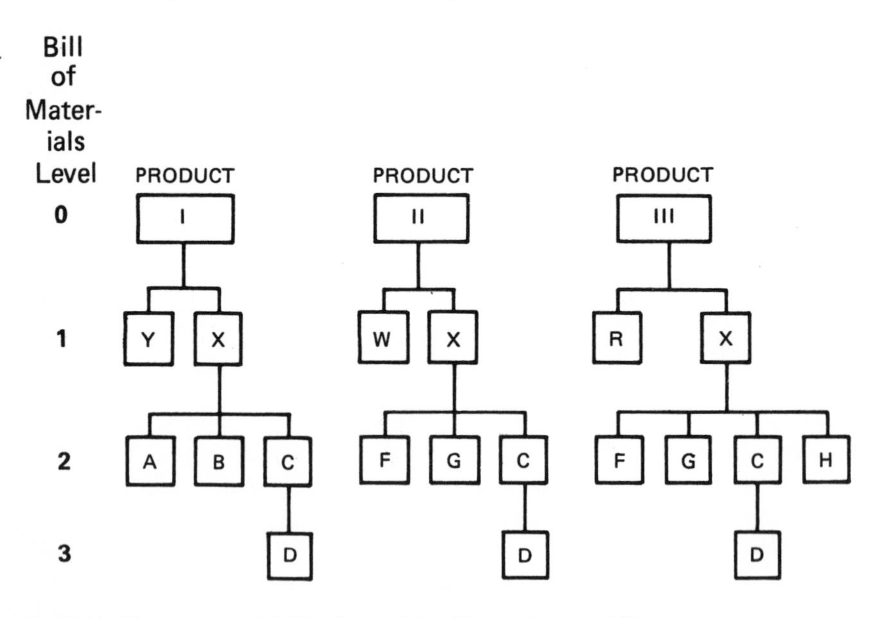

Fig. 2-14. Three product's bills of materials with no phantom bill.

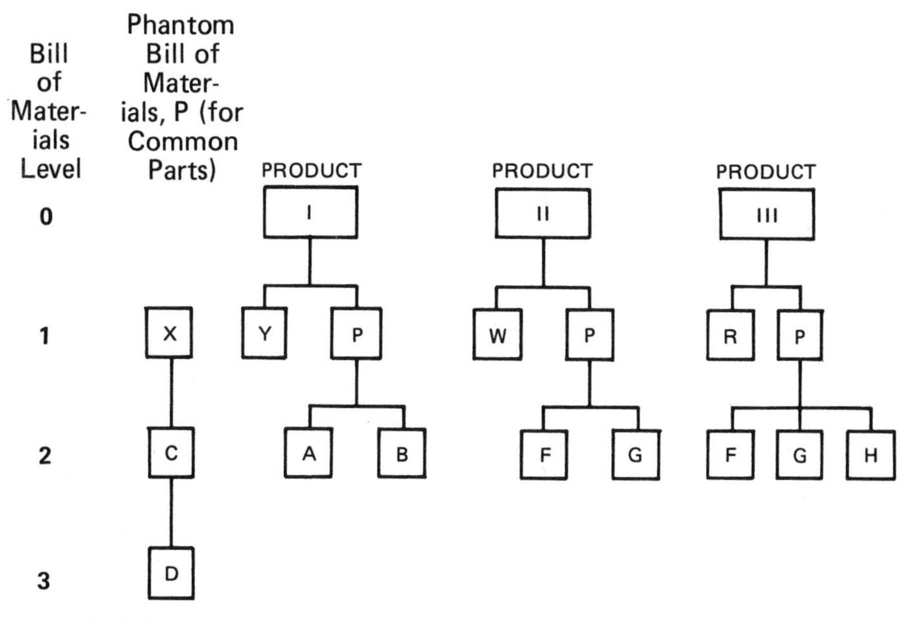

Fig. 2-15. Three product's bills of materials with a phantom bill.

the 10 product bills of materials would show the phantom part number, plus all of the individual parts needed to make up each option. This concept is shown in Figs. 2-14 and 2-15.

Note that the phantom bill of materials does *not* have to be for an end item or a product that is shipped to a customer. It can be for any convenient subgrouping of parts common to many upper-level products.

A Common Bill of Materials Problem

A problem in some companies is that there often is more than one bill of materials for each subassembly or end item. Every function or department in the organization has its own bill of materials for a product. Typically there are five bills of materials for any one product in such a company:

1 / Engineering's—the way the engineers designed it

2 / Cost Accounting's—the way the accountants costed it

3 / Manufacturing's—the way Production Engineering thinks the workers build it

4 / Purchasing's—the way Purchasing buys material for it

5 / Production's—the way the workers *really* put it together

Rarely do all five bills of materials contain the same information regarding part number, description, quantity, effectivity dates, etc. This situation generates confusion and errors in manufacturing operations, and requires five times as much data-base space and maintenance effort than would be required if the entire company agreed that one bill of materials per part was the *one* product the entire manufacturing team concentrated their efforts on.

Ultimately, each part's bill of materials must reflect the way the product is assembled on the shop floor. If the master bill of materials data exist once in the data base, it is perfectly legitimate for different functional departments to *print out* only selected elements of these data on their bill of materials report or even to format their bill of materials report differently. This is a far different matter from having several bill of materials files or data sets. A single bill of materials—*monobomism*—based on the concurrence of Engineering, Manufacturing, Accounting, Purchasing, and Marketing, should be the goal of any well-run manufacturing operations.

How to Achieve Monobomism

Central to the establishment and maintenance of a single bill of materials per product is the concept of having a central department that administers all bills of materials for the company. This department maintains and controls the bill of materials data base, and might report to the Vice President of Manufacturing. Its charter is simple:

> To create and maintain *one correct* and *complete* bill of materials for each end item the company sells. Bill of materials structuring should be coordinated among user functions and bill of materials data definitions communicated to all users. It is important that this bill of materials design process incorporate the needs of other users such as Accounting and Engineering.
>
> To maintain all bills of materials effectivity dates accurately, including new- and/or prototype-part bill of materials. In addition, obsolete-part bill of materials must be periodically purged from the active data base to a historical bill of materials tape.
>
> To maintain an audit trail of all design and part specification changes.
>
> To create phantom bills of materials where possible that will structure products for manufacturing or inventory more efficiently.

Input to the bill of materials control organization should come from a committee that includes members from Engineering, Manufacturing (not only management but assembly line workers), Finance, Purchasing, Quality Control, Marketing, and Field Service (i.e., replacement parts and service). Changes to bills of materials should be evaluated by this committee with regard to timing, cost effectiveness, customer appeal, product liability considerations, product life cycle, etc. A single bill of materials for each end product will focus all efforts toward getting and keeping each bill of materials 100% accurate. In addition, the company will then be manufacturing its products from one common base.

3

Manufacturing Process and Related Manufacturing Data Requirements

In this chapter other manufacturing data concepts such as item master listing, parts/raw material data base, routing, and bills of labor will be examined. The last two of these four items form the basis for shop floor control and capacity planning.

Routings or Process Sheets

In examining the way a product is made, most production control or manufacturing people divide the manufacturing process into a series of major steps or stages of production called work centers, rooms, departments, or assembly lines. Operations of a similar nature are grouped within these areas. Inside these major work center areas, the work can be further divided into work stations or machine centers. In order to establish standards and/or to track manufactured items through the factory, a way is needed to locate each item being worked on. Also, a way is needed to identify, even on a "Macro" level if necessary, the operations through which the work must pass. The process sheet, or routing, lists in sequence each manufacturing operation a part must go through to be manufactured. This operation sequence and location concept is illustrated in Fig. 3-1.

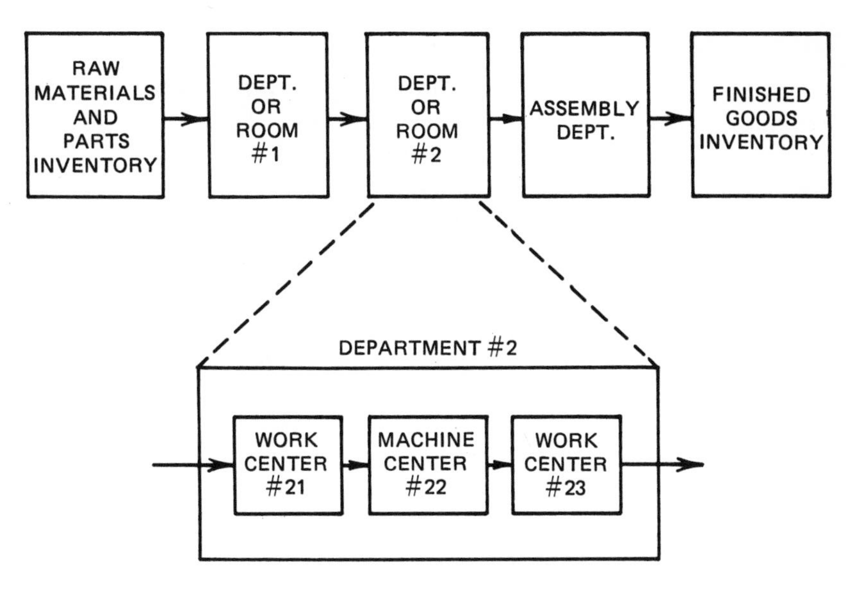

Fig. 3-1. Stages of production.

Given only the information in this figure, we can begin to establish a routing for rough cut capacity planning by listing the work centers and putting in some average production rates or times for each department. These times could come from a statistical study performed over some production period, or from standard time study figures, or just the foremen's best estimate. (See Fig. 3-2.)

Department number	Department	Production cycle time
100	Machining	2 hours
200	Assembly	3 hours
300	Crating	1 hour
		Total 6 hours

Fig. 3-2. A simple routing or bill of labor for part #1050.

We now know that part #1050 takes 6 hours to produce. From this base, we can deduce even more information. In a 7.5 work hour shift we can make 1.25 parts. In five one-shift days we can assemble 6.25 parts; or in a typical 250 day year, we can make roughly 313 parts. Thus, we very quickly arrive at a rough cut capacity plan!

Labor Hours

The 6 hours it takes to manufacture one part #1050 is a production cycle time. It tells us nothing about the number of labor hours required to produce the part. The number of labor hours required depends on how many people are at each work center, work station, or machine center. We can expand the routing to show more detail (see Fig. 3-3).

Department	Labor hours required	Production cycle times
100 Machining	12	2 hours
200 Assembly	6	3 hours
300 Crating	2	1 hour
	Total 20	6 hours

Fig. 3-3. An expanded routing for part #1050.

	Labor hours required
Department 100	
Machine center 120	6
Work station 140	4
Machine center 150	2
Subtotal Dept. 100	12
Department 200	
Work station 210	4
Work station 270	2
Subtotal Dept. 200	6
Department 300	
Work station 350	2
Subtotal Dept. 300	2
Total	20

Fig. 3-4. A more-detailed routing for part #1050.

We now know that each part #1050 requires an input of 20 standard labor hours to manufacture and crate it for shipment. Therefore, if we want to make 313 units/year, we need to staff and pay for 313 × 20 = 6260 man hours of work. Dividing by 250 work days/year, 25 man hours/work day would be required.

Now, we can begin to move from the macro to the micro by setting up the routing to include every work station or machine center (Fig. 3-4).

Operation Cycle Times

How long does each part wait to be worked on? How long does it take to move parts between work centers? Is there any machine setup time, etc.? I want to digress and discuss the production cycle time.

Typical work center operation cycle times are made up of the elements shown in Fig. 3-5. (The noun *queue* is the English equivalent of a waiting line, such as a line of people waiting for something to happen.) Referring to Fig. 3-5 we have

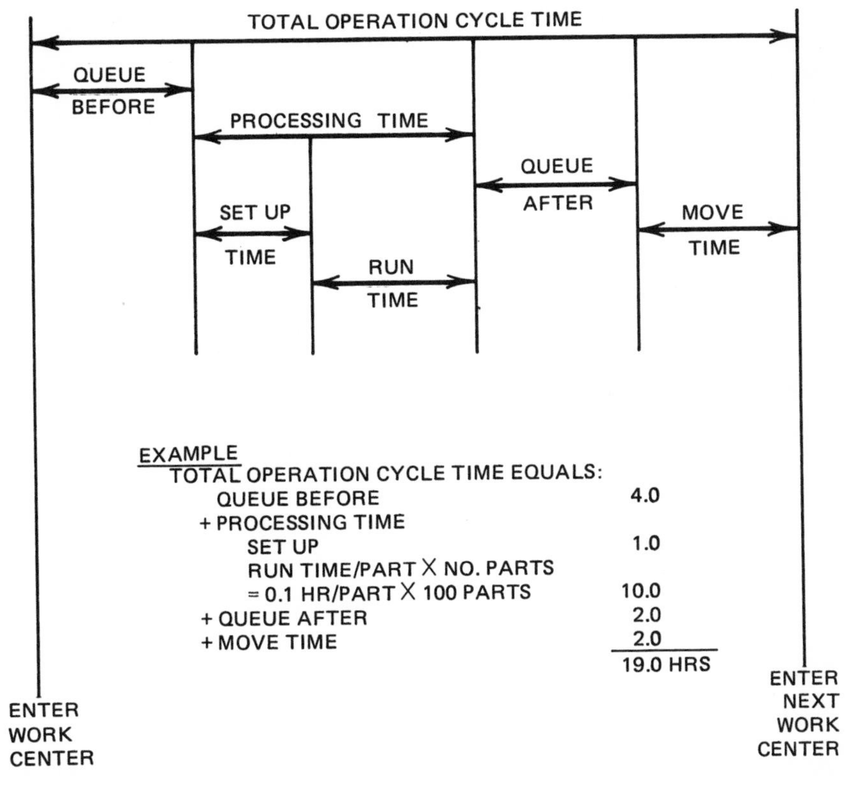

Fig. 3-5. Elements of total operation cycle time.

Queue Before

Wait for the part to be put in the machine or up on the work bench.

Processing Time

1. Setup
   Setup on a machine, in clamps, fixtures, etc.
2. Run Time
   Actual work on the product—i.e., machining operations, welding, etc. This is calculated by multiplying the time per part times the lot size (number of parts).

Queue After

Wait for the part to be moved to the next work center.

Part #1050
Labor Hours Required

	Setup	Machine run	Assembly	Total
Department 100				
MC 120	2	4		6
WS 140			4	4
MC 150	1	1		2
Subtotal	3	5	4	12
Department 200				
WS 210			4	4
WS 270			2	2
Subtotal			6	6
Department 300				
WS 350			2	2
Subtotal			2	2
Total	3	5	12	20

Fig. 3-6. Part #1050's routing with setup, run, and assembly times added. (MC, machining center; WS, work station.)

Move Time

Actual move time to transport the parts to the next work center.

Many studies show queue time alone often takes 80–90% of the entire production cycle time, especially in job shop environments. In fact, normal run or "up" time in job shop environment for a given part is usually only 5% of total available work in process (WIP) time!

So far, we have assumed that the example routing labor times have included all the components of the total operation time. Assuming the work stations are close to each other, and that we have an ideal environment where there is no queue time, we then might find the situation illustrated in Fig. 3-6.

Now, we can begin to see the benefits of detailing the labor required. Different skills and pay rates are required for different shop floor operations. In addition, perhaps, different supervisory skills or knowledge may be required in different departments.

Craft Codes and Labor Costing

Relating the labor hours shown above to a register of craft skills so that we can obtain manpower requirements by hours and skill type or level, the final routing might look like Fig. 3-7. The craft code reference table this routing is based on is shown in Fig. 3-8.

Routing for Part #1050
Standard Labor Hours Required

	Machine				Assembly				
	Craft code	Setup Time	Craft code	Run Time	Craft code	Setup Time	Craft code	Run Time	Total Time
Department 100									
MC 120	35	2	60	4					6
WS 140					10	2	20	2	4
MC 150	35	1	60	1		—		—	2
Subtotal		3		5		2		2	12
Department 200									
WS 210							20	4	4
WS 270							20	2	2
Subtotal								6	6

Fig. 3-7. Final routing for part #1050.

Department 300						
WS 350			10	2		2
Subtotal				2		2
Total	3	5		4	8	20

Fig. 3-7 (*continued*)

Craft code	Description	Standard pay rate/hour
10	Unskilled laborer	3.50
20	Semiskilled assembler	4.60
35	Semiskilled setup machinist	5.00
60	Skilled machinist	8.00

Fig. 3-8. Craft code reference table for part #1050's routing.

Now, we can establish on a standard cost basis exactly how much any quantity of part #1050 will cost to manufacture, and the number and skill levels of the people that we need to employ to make this product. The standard labor cost for part #1050 can be calculated as shown in Fig. 3-9.

Craft code	Number of hours	@	Rate ($/hour)	=	Labor cost ($)
35	3.0		5.00		15.00
60	5.0		8.00		40.00
10	4.0		3.50		14.00
20	8.0		4.60		36.80

Total labor for one part #1050 = $105.80

Fig. 3-9. Standard labor cost for part #1050.

Routings

The worker out on the shop floor needs instructions on how to make or assemble the product. These instructions can be combined with other information to create a routing or process sheet—a step-by-step sequence and set of instructions for making the product. For our part #1050, this routing might look like the one illustrated in Fig. 3-10.

Department	WS	MC	Description	Craft code	Standard labor hours
100		120	Setup casting	35	2.0
		120	Drill 2 each $\frac{1}{2}''$ diameter holes	60	0.5
			Tap holes, $\frac{1}{2}'' \times 20$ NC	60	0.5
			Mill top surface	60	1.0
			Counterbore bearing hole	60	2.0
	140		Assemble bearing housing	10	2.0
			Attach subassembly #6	20	2.0
		150	Setup assembly	35	1.0
		150	Drill 3 each $\frac{1}{4}'' \times 2''$ diameter holes	60	0.3
			Ream 3 each $\frac{1}{4}'' \times 2''$ diameter holes	60	0.2
			Install locating pins	60	0.5
200	etc.				

Fig. 3-10. Routing with instruction for part #1050. (MC, machining center; WS, work station.)

Depending on data-base organization and the level of complexity desired, one might want to create only routings for rough cut capacity planning, or craft code data and routings, or combine all this information on one report that shows a company's manufacturing process for each part.

The point is that these data can convey such vital information regarding work centers, skill levels, craft types, hours required (both labor and cycle time), and labor costs. This information allows one to *control* work on the shop floor (is it at the right place at the right time?), and *plan* future capacity by work center, machine type, craft skill, number of people, standard cost, etc. We will discuss capacity planning in Chapter 6 and shop floor control in Chapter 7.

Materials Data

Materials-related data have to exist in addition to the bill of materials and process-related data. This information is usually summarized on the material or item master report and the raw material inventory report. Sometimes, these two reports are combined as one. Other data reports related to the purchasing of raw materials or parts will be discussed in Chapter 8.

Material Master Report

The material master report is a complete listing of every material used by the company in manufacturing its goods. It can show such information as

1 / Part number
2 / Long description
3 / Short description
4 / Unit of measure
5 / Status code
6 / Effectivity dates
7 / Buyer code
8 / Part standard cost

1 / Part Number. A company-assigned part or product number—usually broken down by product *class* or family, e.g., fasteners, then by individual description. For example,

Class 110=fasteners
120=bearings
130=adhesives

part #110-121953 nut
#110-061706 bolt

2 / Long Description. A "long" description (30–45 spaces) identifies the product as fully as possible. This description might be:

bolt- $\frac{3}{8} \times 24 \times 1\frac{1}{2}$, grade 8, chrome plated

3 / Short Description. A "short" description of 12–20 spaces is used when descriptions are printed on stockroom tickets, work orders, etc., where the full description would take up too much room. For example,

bolt-$\frac{3''}{8}\times 24\times 1\frac{1''}{2}$-GR8-CRP

4 / Unit of Measure. The unit of measure, e.g., pounds, gross, dozen, pair, inches, feet, yards, or gallons. These units of measure must be agreed upon at the outset of manufacturing data creation, and published as a selection list of possible units. Any changes to this list should be made by the bill of materials (BOM) committee.

Sometimes there are differences between units when parts or raw materials are *bought* and as they are *used* in the manufactured product. Naturally, whatever unit of measure is used makes a great difference to the price that is associated with it. For instance, bolts may be bought by the box of 100 and used as one each. It is my opinion that the unit of measure on the material master should be the lowest unit of measure when the part is utilized. Sometimes this requires prices carried out to three or four decimal places, but that is a small price to pay (no pun intended) to avoid the confusion of having the price cover a varying purchase quantity, or supplier package/pricing quantity.

5 / Status Code. A status code can reflect such things as *N—New:* not in use yet—listed on the data base but not active. *A—Active:* in use or available for use. *O—Obsolete:* no longer available for use but must be maintained on data base until the remaining inventory is scrapped or sold. *D—Deleted:* may be purged from the data base—product number no longer valid and material no longer exists.

6 / Effectivity Dates. Effectivity dates—may also be shown here in addition to the BOM—show date item entered, the latest maintenance date (for an audit trail covering material master data item changes).

7 / Buyers Code. There might be a buyer code showing the buyer for each item.

8 / Part Standard Cost. Standard cost can be shown also, if the company has such an accounting system. Note that this is different from the purchasing data or even inventory data that may show vendor prices or the historical cost picture for a product.

Actually, the material master data can include *anything* that a company considers relevant to each raw material item. The descriptions listed are fairly standard, but that is not to say that there are not more or different terms for a manufacturing application.

Raw Material and Part Inventory Data

These data convey inventory status information for all raw materials or stocked parts, subassemblies, tools, or supplies. Typical data carried here are

- Part number
- Long and/or short description
- Unit of measure
- Whether the part is made or bought
- Buyer code
- Standard cost
- Vendor codes—i.e., supplier names or codes for all authorized sources
- Latest prices for each vendor
- Lead time for each vendor
- Lowest level of bill of materials where part is used
- Part location—e.g., warehouse, stockroom, row, bin number, or incoming inspection or receiving dock
- Re-order point quantity—if not mrp system
- Safety stock quantity
- Standard lot size—quantity for each order—can apply whether mrp or not
- Quantity on order and from which vendor(s), purchase order number (if not mrp system)
- Issues—per week, month, month to date, year to date
- Receipts—per week, month, month to date, year to date
- Inventory check procedure code—e.g., inventory by min/max, or weighing

Other types of data can be saved in the area of the company's data base associated with inventory data at the discretion of the Data Base Administrator. Reports can be generated from these data such as the inventory stock status report.

The basic data needed to run a manufacturing planning and control system are now in place. Now we can examine how to effectively utilize these data to plan and run a manufacturing operation.

4

Material Requirements Planning

Until recently, material requirements planning was known as MRP. Lately, MRP has come to stand for manufacturing resource planning—now "Big MRP." Manufacturing resource planning refers to the entire closed-loop modern manufacturing planning and control system that includes forecasting, master scheduling, capacity planning, material requirements planning, shop floor control, etc. Today, "Little MRP" is used to denote only the material requirements planning module of the manufacturing planning and control system. A reminder to readers—in this book, mrp is material requirements planning; MRP is manufacturing resource planning.

The Concept of Materials Control

In looking at the materials aspect of a manufacturing control system, we are interested in answering the question: What factors concerning materials do we want to control? Answer: The basic function of material control is to have enough of the right materials available at the right time to manufacture the company's products.

Note that there are three fundamental things this statement says we need to control, namely,

1 / the right *parts*

2 / the right *quantities*

3 / the right *timing*

Given the first two items, the proper timing is a matter of *priorities.* Which part do you want first, second, etc.? Material requirements planning is primarily a *priority* planning or scheduling tool—it *plans* the company's manufacturing priorities by creating a schedule; and with proper feedback from Purchasing, Inventory Control, and the shop floor, mrp controls those scheduled *priorities.* The planning and control of these priorities is designed to ensure that the company's production schedule is met, and either inventory levels or customer shipments are those desired by top management.

Dependent-Parts Demand

Material requirements planning is based on the concept of dependent-part demand. Here, we take advantage of the fact that most manufactured products are arranged in some hierarchical structure of subassemblies and/or component parts. In the language introduced earlier, most products have several levels in their bills of materials where the number of lower-level parts needed depends exactly on the requirements for higher-level parts.

If for any product with more than two levels to its bill of materials, for example, we want to manufacture 10 parent items, then the demand for the component parts is *dependent* on the demand of the end items, in this case 10 end items. It is assumed, of course, that the bill of materials is correctly structured and accurate. This concept is illustrated in Fig. 4-1 for a set of pruning shears.

Part #	Description	Quantity
979, pruning shears		
078	Blade, LH	1
088	Blade, RH	1
101	Bolt, $\frac{3}{8}''$ ×24 ×1″	1
206	Nut, self-lock, $\frac{3}{8}''$ ×24	1
427	Handle, wood, 27″	2

Fig. 4-1. Bill of materials for part #979, pruning shears.

Therefore, to make 10 #979's, we would need *exactly* 10 078's, 10 088's, 10 101's, 10 206's, and 20 427's. This, of course, assumes no goods would be scrapped or damaged in production, that there would be no quality-control rejects, and that no shears were in stock or in process. Note that the component or the dependent demand is an *exact* by-product of the end item requirement. Therefore, we only have to forecast or state each end item that will be built or stocked for the customer—not each and every lower-level part we stock.

The initial explosion of the parts required to support a master production schedule results in what are called *gross* requirements. Against these gross requirements, we then have to subtract the parts we already have on hand or on order to arrive at the *net* requirements for which new orders must be issued. This simple *netting* process is illustrated in Fig. 4-2 for the #078 blade used in the preceding explosion.

Gross requirements	10
On hand	3
	7
On order	4
Net requirements	3

Fig. 4-2. A simple netting process for part #078.

In a broader sense, whether dependent demand follows a vertical structure depends on the viewer's frame of reference. If, for instance, a piece of sales literature or a warranty card is shipped with each finished good, this would be a horizontal bill of materials item in a *manufacturing* sense but vertical from the shipping department's point of view. The final shipped product could for all practical purposes contain a *shipping* bill of materials such as the one in Fig. 4-3.

XXX	Steam iron assembly
YYY	Box
ZZZ	Warranty card
PPP	Instruction manual
QQQ	Sales literature (other products)

Fig. 4-3. A shipping bill of materials for an XXX steam iron.

Independent-Parts Demand

Independent demand refers to demand that is independent of any other inventory item. This demand is usually random and must be forcast in some manner—whether by a guess or by using a sophisticated statistical forecasting method. Most finished-goods demand is independent, since ordering one size and style of shoe, for instance, has no effect on any other shoe in the warehouse.

Re-Order Point Inventory Control

The benefits of mrp can be seen by first examining how the old re-order point statistical inventory control was supposed to work. Essentially, re-order point inventory control was built around the concept of keeping *extra* inventory on hand in the form of safety stock to cover unforecasted demand, vendor supply problems, quality rejects, record keeping errors, lost parts, and part usage during the part-replenishment lead time.

The basic idea of the order point technique was to work inventory down to the re-order point and then issue a purchase order for a quantity of parts based on some lot-sizing technique. The order quantity (lot size) was calculated based on expected part demand during the vendor's lead time to deliver the part order and the service level the company desired to maintain over this period. The concept is illustrated in Fig. 4-4.

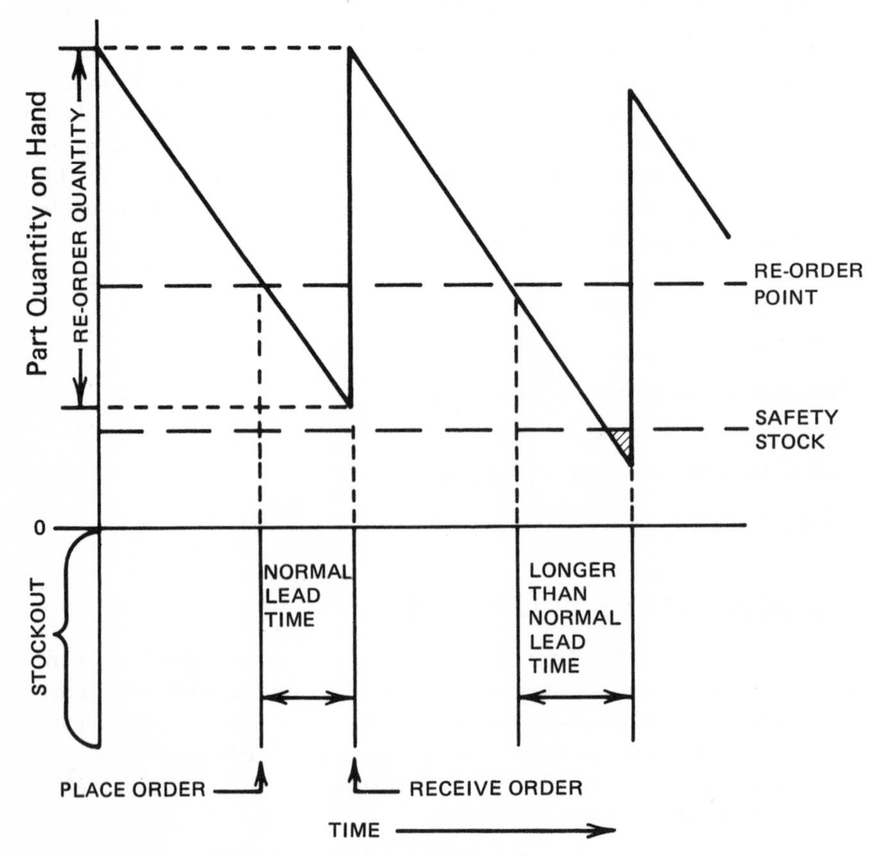

Fig. 4-4. Diagram of re-order point inventory control.

Note that little in the re-order point method is tied to the idea of time-phased need, that is, having the right quantity available at a specific time to make a certain number of end items. Instead, the idea was to *always* have enough stock on hand to cover any part demand situation that would be likely to occur.

As an inventory control method in the discrete-product manufacturing environment, the order point technique is flawed for several reasons:

1 / It is individual-part-based and not focused on the demand of the parent item that the part goes into.

2 / It is based on historical usage of the part—not the future manufacturing requirements for the product(s) in which the part is used. More specifically, it also assumes that future part or product demand will be similar to past demand. This may not be true if a product is discontinued or undergoes an engineering change, for instance.

3 / The order point method, with its allowance for safety stock, creates higher levels of inventory than are needed for future manufacturing requirements. This results in higher inventory holding costs due to extra storage space requirements, extra insurance needs, and extra material handling. Also, of course, it ties up the company's cash in unnecessary inventory investment.

4 / The order point method also assumes a constant part usage rate (indicated by the slope of the demand line). It is a rare company where this smooth demand occurs—especially in a discrete-product environment as opposed to a process-oriented industry.

5 / Above all, the order point method is completely insensitive to the *timing* of future part requirements. Instead, the order point method concentrates on quantity and *when* to *order,* not *when* and *if* the part is needed in the *future.*

Two other easily refuted factors concerning order point techniques are often cited as reasons to no longer utilize the method. The first of these arguments is: Imagine following the re-order point process for 20,000 individual inventory items or more—it would be impossible! Today, computers can easily handle that situation. In fact, computers would have a far easier time calculating safety stock, re-order points, and service levels than calculating mrp requirements.

Second, flaws in the most commonly used economic-order-quantity (EOQ) lot-sizing technique are often cited as arguments against the order point techniques. This is a specious argument, however. Lot sizing still has to be accomplished by some method in either an order point or an mrp system. Criticisms may be just as valid against EOQ-based lot sizing for mrp as they are for re-order point systems.

In any case, the previous two "flaws" of order point techniques really do not apply. The five previously cited objections are the main drawbacks to the old re-order point approach.

In summary, the order point technique has the following characteristics:

It is a part-by-part approach.

It is based on historical part demand.

It is not oriented toward time-phased end item product need.

It results in higher than necessary inventory levels.

We will now discuss how mrp can improve this situation.

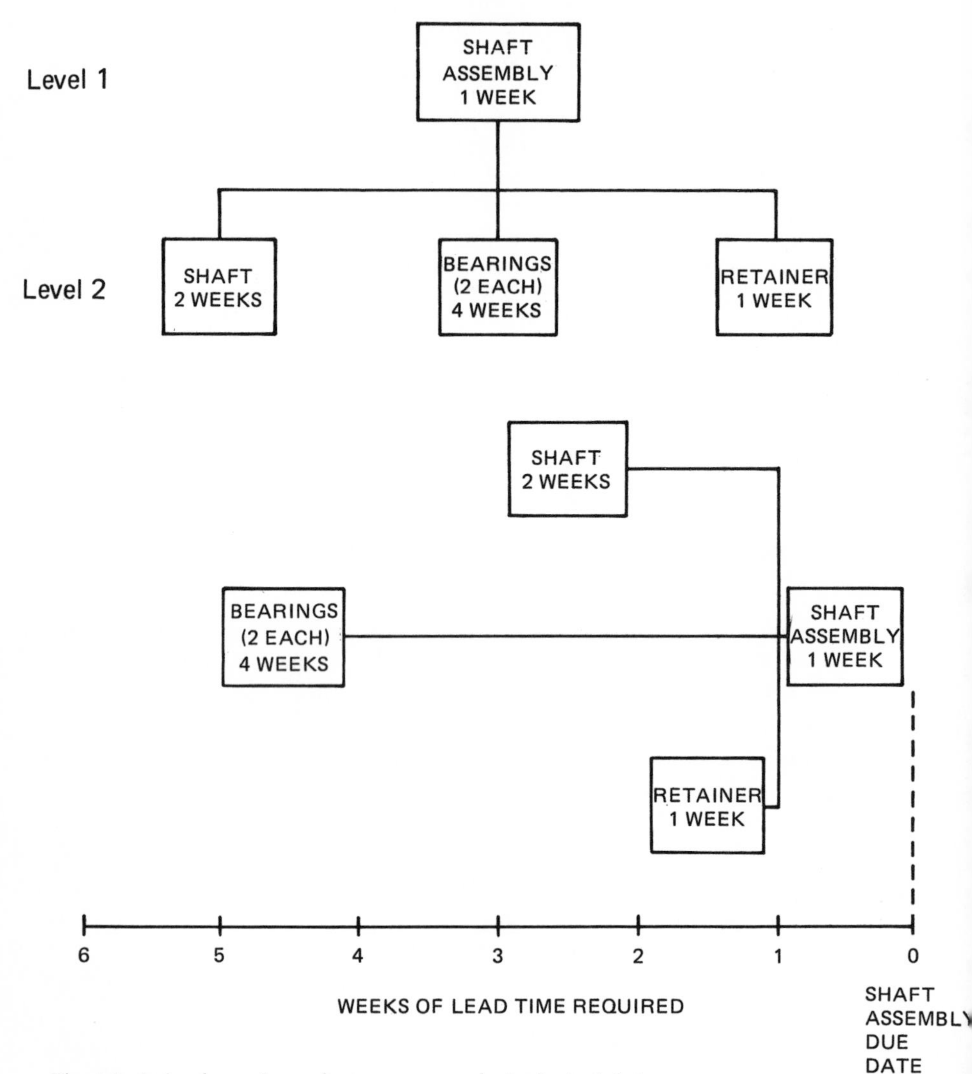

Fig. 4-5. A simple part's product structure to be backscheduled.

The mrp Scheduling Process

The central feature of mrp is the ability to precisely calculate and maintain the *due date* (or *need date*) of dependent parts based on the due date of the order for the parent or end item. This process is called backscheduling from the due date. This backscheduling offsets the start date for a part by the lead time required to produce the part in time to meet the due date of the next-higher-level part. A simple example of backscheduling is shown in Fig. 4-6 for a two-level part whose product structure and lead times are shown in Fig. 4-5.

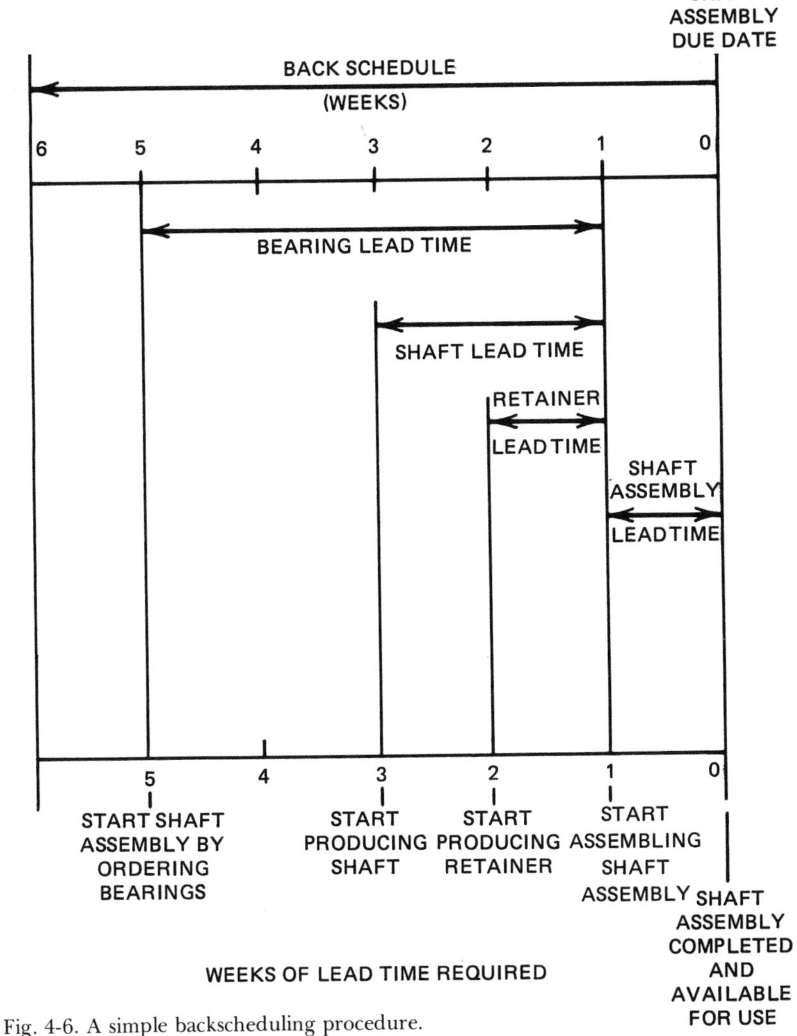

Fig. 4-6. A simple backscheduling procedure.

Note in this example that the procurement or manufacture of each part is given a priority so that all parts are timed to support the due date of the shaft assembly. No part is made or procured and unnecessarily stocked before it is needed.

For backscheduling to work properly, we need to have accurate information on the time required to produce or procure every component or assembly. For manufactured parts, this information is based on time standards in the part's routing data. (For purchased parts, the lead time to procure parts, known as the vendor lead time, is maintained for each part by the purchasing function.)

Using the part routing shown in Chapter 3 and repeated in Fig. 4-7, we will see how mrp logic schedules an order for a manufactured part to the shop floor. For convenience we will assume 2 hours move time between departments and 4 hours queue time at each operation. We also need to know how many hours each machining center (MC), work station (WS), or department is *available* to perform work on a part. This

Routing for Part #1050
Labor Hours Required

	Setup	Run	Assembly	Total
Department 100				
MC 120	2	4		6
WS 140			4	4
MC 150	1	1		2
Subtotal	3	5	4	12
Department 200				
WS 210			4	4
WS 270			2	2
Subtotal			6	6
Department 300				
WS 350			2	2
Subtotal			2	2
Total	3	5	12	20

Total labor hours required = 20

Fig. 4-7. Routing for part #1050. (MC, machining center; WS, work station.)

available-hours figure can be different for different areas of the plant. In addition, after knowing how many shifts each area or department works, some allowance must be made for worker breaks and other lost worker time. For illustrative purposes assume Department 100 works three shifts, Department 200 works three shifts, and Department 300 works two shifts; and allow $7\frac{1}{2}$ hours productive time per shift. In addition, assume a lot size of 250 parts.

For each stage of the manufacturing process, we need to calculate how many hours the part will be in that stage, and then convert those hours into work days of production weeks. Total manufacturing time in hours will equal move time plus queue time plus setup time plus the lot size multiplied by the run time per part. Therefore,

in Department 100

in MC 120 — 2 hours setup + 250 parts × 4 hours run/part = 1002 hours required

+ queue time before WS 140 = 4 hours

in WS 140 — 4 hours assembly time/part × 250 parts = 1000 hours required

+ queue time before MC 150 = 4 hours

in MC 150 — 1 hour setup + 250 parts × 1 hour run/part = 251 hours required

+ 2 hours move time to Department 200

in Department 200 — + queue time before WS 210 = 4 hours

in WS 210 — 250 parts × 4 hours assembly/part = 1000 hours required

+ queue time before WS 270 = 4 hours

in WS 270 — 250 parts × 2 hours assembly/part = 500 hours required

+ 2 hours move time to Department 300

in Department 300 — + queue time before WS 350 = 4 hours

in WS 350 — 250 parts 2 hours assembly/part = 500 hours required.

These totals are illustrated in Fig. 4-8.

Department	Setup and run	Queue	Move	Total
100	2253	8	2	2263
200	1500	8	2	1510
300	500	4	—	504
Total	4253	20	4	4277

Fig. 4-8. Total hours required for 250 part #1050's.

Now further assume that there are three men/shift in Department 100, two men/shift in Department 200, and one man/shift in Department 300. In addition, we have three machines each in Department 100, MC 120, and MC 150. The hours available/day are then calculated by

Department 100
3 men/shift × 3 shift/day × 7.5 work hours/man
= 67.5 work hours/day.

Department 200
2 men/shift × 3 shift/day × 7.5 work hours/man
= 45.0 work hours/day.

Department 300
1 man/shift × 2 shift/day × 7.5 work hours/man
= 15.0 work hours/day.

Therefore, the number of days needed in

$$\text{Department 100} = \frac{2263 \quad \text{hours}}{67.5 \quad \text{hours/day}} = 33.5 \text{ days}$$

$$\text{Department 200} = \frac{1510 \quad \text{hours}}{45.0 \quad \text{hours/day}} = 33.6 \text{ days}$$

$$\text{Department 300} = \frac{504 \quad \text{hours}}{15.0 \quad \text{hours/day}} = 33.6 \text{ days}$$

The total manufacturing time is then 100.7 days.

In this simple example we have ignored the stock withdrawal or pick out time necessary to get the parts or raw materials to Department 100, and the move time required to take the finished subassembly from Department 300 to the stock room.

The mrp logic now rounds off the 100.7 days to 101 days and backs up 101 manufacturing days from the order's due date. At this point it would schedule ("slot") the work to begin on day "due date minus 101". If we are dealing in weekly time buckets, the order will be rounded off to start in the week where day "due date minus 101" resides, namely, in the 21st week before the due date if no holidays are counted.

This step-by-step backscheduling calculation allows one to know at any given time where in the manufacturing process a part is supposed to be (if it is on schedule). This knowledge forms the basis for shop floor control to be covered in Chapter 7.

It should be noted at this point that we have worked through an example of mrp backscheduling that is theoretically correct and available for use in at least two state-of-the-art software packages. In earlier, less-sophisticated mrp software, a fixed manufacturing lead time for each part was input and maintained by the planners. This figure was usually carried in the item master or bill of materials data record. Presumably, this fixed manufacturing lead time was based on the routing standards, some assumed move and queue times, and some average lot size.

There are two factors which cause this fixed lead time method to be less than correct. The first is that the assumed lot size may be different from a lot size arrived at by the mrp netting and lot-sizing algorithm (especially in a lot-for-lot environment). This can introduce a sizable distortion in the manufacturing lead time used by mrp if the run time per piece is significant. Second, the manufacturing lead time is further distorted in this fixed lead time offset method since the manufacturing lead time calculation does not take into account the hours available (one, two, or three shifts) of each work center on a part's routing. Usually, some average availability figure is assumed for all the work centers in the

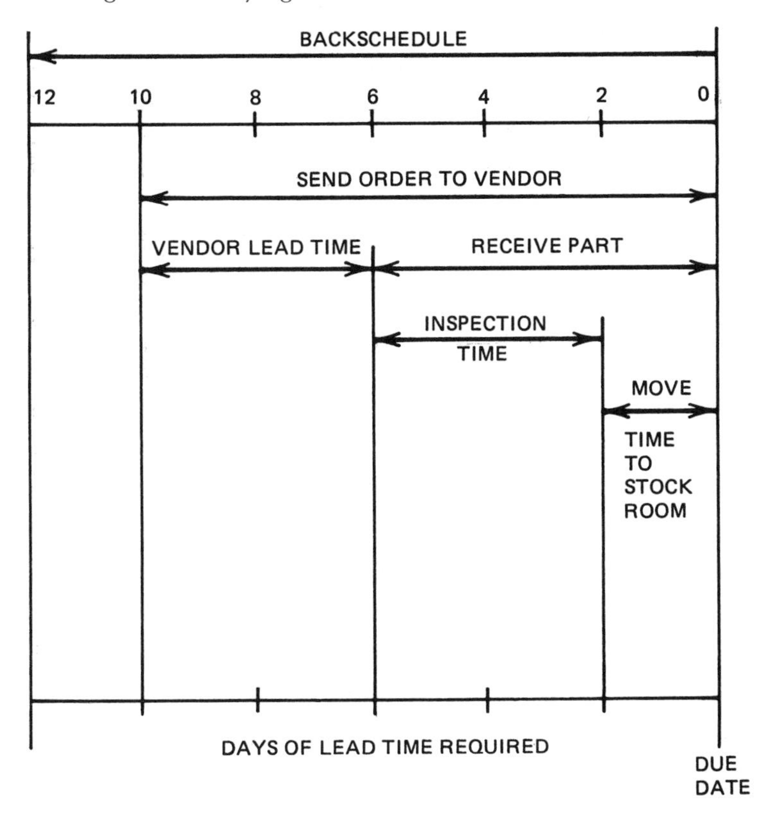

Fig. 4-9. Backscheduling of a purchased part.

entire factory. This does not allow accurate backscheduling either for mrp or for capacity requirements planning purposes.

Manufacturing lead time is strictly a function of lot size (determined by mrp based on the lot-sizing algorithm assigned), manufacturing engineering standards for setup and run time per operation, and move and queue times assigned by production control after work center analysis. Why not let the computer calculate the correct lead time for each planned order as we have demonstrated, rather than place an approximate lead time in the mrp backscheduling algorithm that will result in wrong priorities and excessive rescheduling by future mrp regenerations? In this manner, the same backscheduling algorithm that plans mrp orders will be used in capacity requirements planning.

We have examined the scheduling of a shop order for a manufactured part. If the part was *purchased,* the vendor lead time would become the basis for scheduling the order. In addition, however, we might like to detail more transactions at this point, and include separate processing times for receiving, incoming inspection, and the movement of parts to the stock room as shown in Fig. 4-9.

mrp Report Formats

Since material requirements planning is time related, a way of showing "time-phased" events occurring has to be adapted. The accepted format for this is a series of time periods or "buckets":

Time Period

0 1 2 3 4 5 6

In bucketless mrp systems, each shop work day is shown by a shop calendar number, usually related to the production day's date.

In bucketed mrp systems, daily requirements are aggregated into buckets of time, which can be either weeks, months, quarters, or years. Weeks are most commonly used for manufacturing planning purposes in a regenerative mrp system, days are most commonly used in a net change system. Note that in whatever time period is used manufacturing management must establish a policy for when during the time period an activity is scheduled or defined to occur. For instance, if weekly buckets are used, Manufacturing will want all their material on Monday morning of a week. Purchasing/Inventory Control will assume they have until Friday afternoon to get the material in the factory and/or out on the shop floor. Some compromise between these two approaches must be agreed upon.

It is also important for manufacturing to work with other management in the company to establish a common company calendar of production days. This gets everyone marching to the same drummer. A common schedule used is to make each quarter 13 weeks long. Then Month 1 of the quarter is always the first 4 weeks, Month 2 the next 4 weeks, and Month 3 the last 5 weeks. Some such quarterly schedule of weeks should be agreed upon so that the entire company works to a common schedule. In this way, fiscal and shipment and production weeks or months or quarters all end on a common date, and no time is wasted juggling figures or doing month-end or quarter-end closings in the middle of a production week. Many companies do not have the computer resources to do month-end closings in the middle of a week without in some way disrupting production operations.

A companywide production calendar means a coordinated effort is possible, and there are no mysteries for anyone as to when fiscal or production benchmarks occur. Such a calendar is shown in Fig. 4-10.

	WEEK	S	M	T	W	Th	F	S	QUARTER
JAN	1	30	31	1	2	3	4	5	I
	2	6	7	8	9	10	11	12	
	3	13	14	15	16	17	18	19	
	4	20	21	22	23	24	25	26	
	5	27	28	29	30	31	1	2	
FEB	6	3	4	5	6	7	8	9	
	7	10	11	12	13	14	15	16	
	8	17	18	19	20	21	22	23	
	9	24	25	26	27	28	1	2	
MAR	10	3	4	5	6	7	8	9	
	11	10	11	12	13	14	15	16	
	12	17	18	19	20	21	22	23	
	13	24	25	26	27	28	29	30	
APR	1	31	1	2	3	4	5	6	II
	2	7	8	9	10	11	12	13	
	3	14	15	16	17	18	19	20	

Fig. 4-10. A common company calendar.

A typical mrp report, where the time buckets are organized horizontally, might look like the example in Fig. 4-11.

		Week					
		1	2	3	4	5	6
Gross requirements		6	13	0	9	4	12
Scheduled receipts			10				
On hand	10	4	1	1	−8	−12	−24
Planned order release		0	10	10	10		

Lead Time = 2 weeks Lot Size = 10

Fig. 4-11. A sample mrp report for part #67831, pump assembly.

Gross requirements come from the master production schedule and are the number of units required for shipment or input to finished-goods inventory.

Scheduled receipts are the number of parts "due in" (expected to be received) on open shop or purchase orders.

On hand is the current inventory balance (number of units in stock) for the part.

Planned order release is the mrp planned order that will be generated in each weekly bucket for planner review, modification if necessary, and release. Most mrp systems will not automatically release an order to the shop floor. The mrp output suggests a planned order to the planner. The planner reviews the order and releases it to the shop floor at the proper time, whereupon it becomes a scheduled receipt.

Note the reflection of the lot size and order lead time in the mrp format. The 10 parts needed in week 4 are ordered in week 2 to cover the 2-week lead time requirement. Although only 4 parts are needed in week 5, 10 are planned by the mrp logic due to the lot size being set at 10.

mrp Netting Logic

Previously, a very simple netting logic to go from gross to net requirements was illustrated. Now, the algorithm or process used in mrp netting will be examined.

To do this, the concept of part *allocation* must be discussed first (Fig. 4-12). The allocation process reserves parts in the stock room needed for released mrp orders that have not been picked yet. Allocations are nothing but uncashed requisitions for stock room material and are absolutely

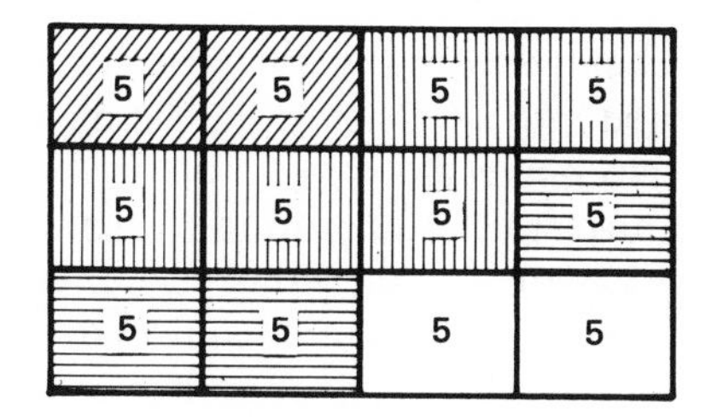

	ON HAND INVENTORY	60
▨	ALLOCATED TO ORDER #582	10
▥	ALLOCATED TO ORDER #639	25
▤	ALLOCATED TO ORDER #695	15
	TOTAL ALLOCATED	50
	TOTAL AVAILABLE FOR NEW PLANNED ORDERS	10

Fig. 4-12. An illustration of part allocation.

necessary for an mrp system to run properly. Allocations prevent parts in stock from being used more than once on paper to fill orders.

To illustrate the mrp netting process, assume in a given time period or bucket a gross requirement for 70 part #27354's. Assume further that there are 40 such parts on order due in in the same period, that the part has a safety stock of 5 each, that 35 such parts are allocated, and that 50 such parts are in stock. The mrp gross to net calculation is shown in Fig. 4-13.

On-hand quantity	50	a
− Safety stock	−5	b
Net on-hand balance	45	$c = a - b$
− Allocated (to unpicked orders)	−35	d
Beginning available balance	10	$e = c - d$
− Gross requirements	−70	f
+ Scheduled receipts	+40	g
= Net requirement	−20	$h = e + g - f$

Fig. 4-13. A sample mrp gross to net calculation.

The net requirement, therefore, is for 20 part #27354's. This requirement would then be fed into a lot-sizing algorithm, and a planned and correctly lot-sized order, corrected by a scrap factor if necessary, would be generated by the mrp system for planner review and release. If the final netting figure had been zero or positive, no net requirement would exist. The existing inventory balance could cover all requirements, and no planned order would be generated by the mrp run.

It is important to note the effect of safety stock and the allocation process on whether or not an order is triggered. If the previous example's gross requirement was only for 45 parts, note what decreasing the safety stock from 5 to 0 does to whether the mrp logic generates a planned order (Fig. 4-14).

	SS=5	SS=0
On-hand quantity	45	45
− Safety stock	− 5	− 0
Net on-hand balance	40	45
− Allocated	−35	−35
Beginning available balance	5	10
− Gross requirements	−50	−50
+ Scheduled receipts	+40	+40
Net requirement	− 5	0

Fig. 4-14. mrp gross to net calculation showing safety stock effect.

Decreasing or eliminating the safety stock results in no order being planned for the time period. Conversely, adding safety stock "pulls in" planned orders on the time horizon, i.e., creates a planned order in this time period where no order was necessary to cover real gross requirements.

Note that in either of the above cases, if mrp is run with no allocation process, its logic thinks that there are plenty of parts on hand to cover the period's gross requirement; mrp then uses the same on-hand parts over and over to fill requirements. In effect, you "push out" planned orders on the time horizon. In reality, the allocated parts are assigned already to previous unpicked released orders and, thus, are *not* available for future planned orders.

Until now we have considered a simple example that showed demand and scheduled receipts both occurring in the same time period. Let us now consider a more real-world example where we have several period's parts requirements and some parts due-in in other time periods.

In Fig. 4-15 we see an example of a regenerative mrp report showing 7 weeks of mrp scheduling for a part. Now, assume there was a change in gross requirements from 0 to 10 units in week 3, from 10 to 15 units in week 5, from 15 to 0 units in week 6, and from 0 to 10 units in week 7. We will follow the calculations needed to arrive at the next week's regenerative mrp report.

Allocated = 10 Lead Time = 2 weeks Safety Stock = 5

		Week						
		1	2	3	4	5	6	7
Gross requirements		20	10	0	0	10	15	0
Scheduled receipts		0	10	0	0	0	0	0
On hand	35	0	0	0	0	−10	−25	−25
Planned order release		0	0	10	15	0	0	0

Assume: Lot-for-lot lot sizing where net requirements equal order quantity.

Fig. 4-15. A real-world mrp report.

The non-time-phased netting process proceeds as before for period 1 (Fig. 4-16). Now the time-phased part of the mrp logic begins where the

On hand	35
−Safety stock	− 5
Net on-hand balance	30
−Allocated	−10
Beginning available balance	20

Fig. 4-16. The non-time-phase netting process.

timing of gross requirements is examined in relationship to that of scheduled receipts:

Period 1 / There are gross requirements for 20 parts in the first time period. The beginning available balance of 20 can cover this requirement and leave 0 available to satisfy period 2's requirements.

Period 2 / Period 2 has 0 parts available, but the 10 due in as a scheduled receipt just cover the gross requirements of period 2 leaving again 0 available for period 3.

Period 3 / Period 3 now has gross requirements for 10 parts but no available balance and no scheduled receipts. In this situation, mrp would replan an order to be released in period 1 (2 week lead time) to cover period 3's requirements.

Period 4 / Period 4's gross requirements are 0, so no planned orders are needed in period 2.

Period 5 / In period 5, there are gross requirements for 15 parts but none available, so a planned order for 15 is created in period 3 to supply period 5's coverage.

Period 6 / In period 6 there are no gross requirements, so the order previously planned for period 4 is no longer necessary.

Period 7 / Period 7's gross requirement of 10 parts will be covered by a planned order for 10 in period 5.

An intermediate mrp report marked to show the rescheduling just done in our example is presented in Fig. 4-17. The new mrp report in its final form after its weekly regeneration is shown in Fig. 4-18.

		Week						
		1	2	3	4	5	6	7
Gross requirements		20	10	10	0	15	0	10
Scheduled receipts		0	10	0	0	0	0	0
On hand	35	0	0	−10	−10	−25	−25	−35
Planned order (old)		0	0	10	15	0	0	0
Planned order (new)		10	0	15	0	10	0	0

Fig. 4-17. Real-world mrp report showing rescheduling (example only—not a part of a regular mrp report).

		Week						
		1	2	3	4	5	6	7
Gross requirements		20	10	10	0	15	0	10
Scheduled receipts		0	10	0	0	0	0	0
On hand	35	0	0	−10	−10	−25	−25	−35
Planned order release		10	0	15	0	10	0	0

Fig. 4-18. Real-world final mrp report.

It is this ability to accommodate change in the manufacturing environment, such as changes in gross requirements, scheduled receipts, and on-hand inventory balances, that makes mrp so powerful as a management tool. As changes occur mrp at each regeneration will automatically recalculate and maintain valid schedule priorities for the thousands of parts that have to be controlled in today's manufacturing environment.

The Use of Safety Stock in an mrp Environment

Generally speaking, there is no place for safety stock in an mrp environment. The use of safety stock is prudent only when there is great uncertainty of *supply* due to international shipment problems or the like, or when there is a very high probability that an entire batch of incoming parts may prove to be almost totally defective in inspection. Safety stock, as we have seen, destroys true mrp priorities by causing the shop floor or Purchasing to be working on orders that are not *really* needed yet. Instead of padding inventory balances with safety stocks, which destroy manufacturing and purchasing priorities, management should concentrate on improving on-hand stock room accuracy and on making sure manufacturing and purchasing lead times are accurate and as short as possible.

Note that creating safety stock to cover *supply* uncertainties is a far different procedure than creating safety stock to cover uncertainties in *demand*—the original purpose of safety stock in the order-point technique. Fluctuations in end item demand are handled by the forecasting and production planning functions—in addition to the finished goods inventory control function if the company is making to stock—in an MRP environment. Once the end items or service replacement parts to be manufactured are master scheduled, demand for lower-level items is an entirely dependent demand, and thus lower-level part requirements are exactly known.

A Multilevel mrp Requirements Explosion

Now, let us examine requirements explosions through several bill of materials levels. All requirements for end items (Level 0's) generate, of course, demands for Level-1 items. Then Level-1 items may generate requirements for Level 2 and so on. This is illustrated in Fig. 4-19. Note that at each level, gross requirements are netted, offset (time-phased) by the appropriate lead time, and lot sized according to the lot-sizing algorithm. Thus planned orders at one level become the next-lower-level's gross requirements.

Level 1—Finished Part

Allocated = 0 Safety stock = 0

Lead time 2 weeks		Week 1	2	3	4	5	6
Gross requirements		6	3	0	9	4	12
Scheduled receipts		0	0	0	0	0	0
On hand	10	4	1	1	−8	−12	−24
Planned order releases		0	8	4	12		

Lot size = lot-for-lot (as required)

Level 2—Machined Casting

Allocated = 0 Safety stock = 0

Lead time 1 week		Week 1	2	3	4	5	6
Gross requirements		0	8	4	12	0	0
Scheduled receipts		0	0	0	0	0	0
On hand	2	2	−6	−10	−22	−22	−22
Planned order releases		6	4	12	0	0	

Lot size = lot-for-lot- (as required)

Level 3—Raw Casting

Allocated = 0 Safety stock = 0

Lead time 2 weeks		Week 1	2	3	4	5	6
Gross requirements		6	4	12	0	0	0
Scheduled receipts		0	0	0	0	0	0
On hand	16	10	6	−6	−6	−6	−6
Planned order releases		20	0	0	0		

Lot Size = 20

Fig. 4-19. A multilevel mrp requirements explosion.

Low-Level Coding

Keeping the above explosion and netting process in mind, a given part number may often exist on different levels of different end items (Fig. 4-20). In the figure part #11 exists as a Level-1 item on part #16, while it is a Level-2 item on part #24. Alternatively, part #11 could be used on two or more levels of the same end item, as in Fig. 4-21.

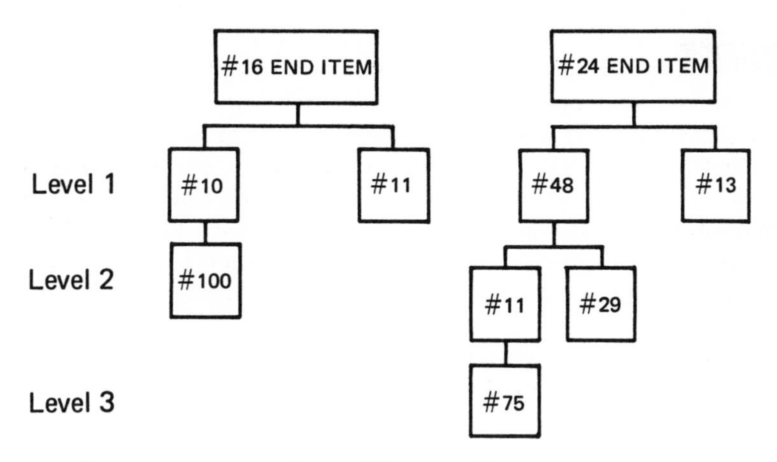

Fig. 4-20. The same part number in two different end items.

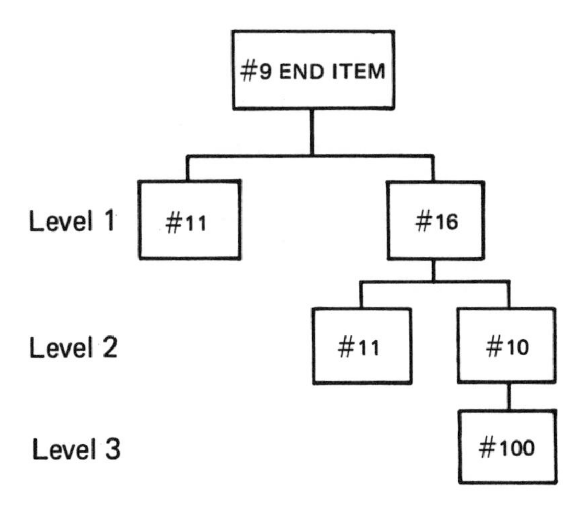

Fig. 4-21. The same part number at two different bill of materials levels.

In the bill of materials explosion process, mrp could handle the requirements calculation by proceeding straight down each end item's bill of materials to the bottom level of the product structure; however, this is cumbersome from a data-processing viewpoint. To reduce the computer's processing time, a low-level code that identifies each part's lowest level of usage in any end item is assigned to all parts or components. Then during the mrp run, totaling the gross requirements of any given item that appears at more than one bill of materials level is postponed until the level designated by the low-level code is reached. The bill of materials for all required items are searched on a level-by-level basis. The first pass through the bill of materials data looks at all Level-1 gross requirements and performs the mrp gross to net calculation for only those parts that have a low-level code of 1. The gross to net calculation for other parts will not occur until each part's low-level code is reached. The second pass looks at all Level-2 requirements, and so on, until the coded lowest level is reached for any required item. Thus, the low-level code is simply a device to tell the computer when to stop in its gross requirements aggregation process for a part and perform the gross to net calculation.

Each level's requirements, when netted and lot sized, become the gross requirements for the next lower level, as we have just seen.

How the Bill of Materials Structure Influences mrp Requirements

In Fig. 4-22 examine the requirements for a simple example, a generator, where that generator is the *only* end item using the listed parts.

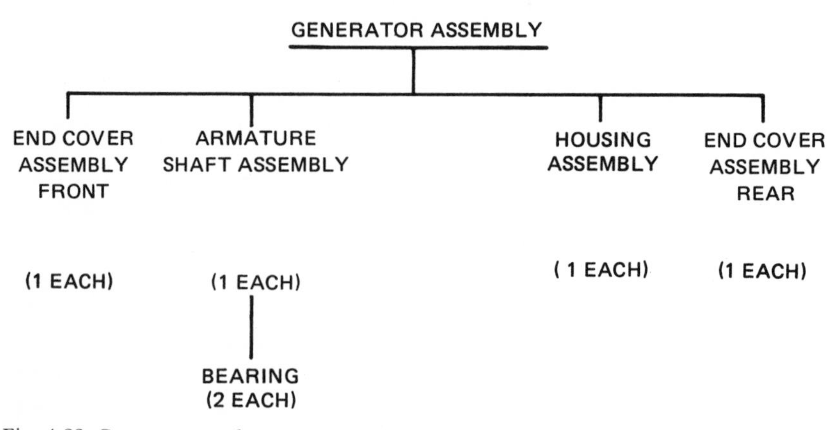

Fig. 4-22. Generator product structure.

Assume a need for 10 generators in week 6 and project the following: On-hand inventories for week 6,

generators	1
armatures	3
bearings	5

Now, how many bearings should be ordered for week 6? The correct answer is *not* 13! A quick, but erroneous, computation leads to 13 by

Generators on hand	1	
−Gross generators required	−10	
Net generators required	− 9	
Bearings on hand	5	
−Gross bearings required	−18	(2/generator × 9 generators)
Net bearings required	−13	

mrp logic nets out these bearing requirements using a level-by-level search through the generator's bill of materials as follows:

Generators on hand	1	
−Gross generators required	−10	
Net generators required	− 9	
Armatures on hand	3	
−Gross armatures required	− 9	(1/generator × 9 generators)
Net armatures required	− 6	
Bearings on hand	5	
−Gross bearings required	−12	(2/generator × 6 generators)
Net bearings required	− 7	

Note that by netting out exact requirments, which reflect the fact that we have a low-level item (bearings) "hiding" in a higher-level subassembly, our requirement for week 6 is reduced from 13 to 7. When you spread this kind of inventory level and purchasing savings through perhaps thousands of bills of materials from 2 to maybe 10 levels deep, you begin to get some idea of the complexity of the mrp netting and scheduling process and why a computer is needed to perform and keep track of all the computations. You also begin to see the inventory dollar *savings* potential of mrp.

Net Change Versus Regenerative mrp Systems

In the early days of computers and mrp systems, most mrp requirements were generated in "batch" computer runs on a weekly basis. Typically, during the week, purchasing and inventory transactions and master production schedule changes were punched out on cards. Then the *entire* mrp report was regenerated in a long computer run over the weekend. This *regenerative* approach completely erased previous planned requirements, and completely replanned and printed out all new planned material requirements for *every* master scheduled item and all dependent parts. This regenerative mrp method resulted in only a weekly "snapshot" of the manufacturing scene.

As data-processing capabilities in both hardware and software developed, someone had the idea that if we change only one end item on the master schedule or one inventory balance, why wait until the weekend to print out *all* the replanned requirements again? Why not print any new requirement changes only for parts affected by master production schedule or inventory balance changes on a daily basis? Thus was born the concept of *net change* mrp. Under this net change concept, mrp-generated ordering and scheduling information can be obtained on a daily basis for only change-affected parts. Because mrp requirements are replanned only for the parts affected by changes—not for every part affected by the master schedule—net change mrp computer runs usually do not take the time that regenerative runs do. The shortened mrp run time is what makes the daily replanning possible. Note, though, that total run time for the week may be as much if not more under net change requirements planning as opposed to regenerative mrp, since the net change program will run five times per week versus the regenerative program's once per week. Periodically, once a month or quarter, a completely new regenerative run can be made, if desired, to provide a fresh starting point for the net change requiremetns planning. The fundamental differences between net change and regenerative mrp systems are now summarized.

Regenerative mrp Systems. All material requirements for master scheduled parts at all bill of materials levels are regenerated once every planning period. Also, all previous mrp planned order output is discarded before the new requirements regeneration. In some manufacturing environments, this regeneration can require 4–20 hours of computer run time. Thus, how frequently a regenerative mrp system runs and material requirements are replanned is often dictated by when data-processing operations can fit the run into its schedule.

Net Change mrp Systems. Net change systems are *transaction* driven by changes to any data considered by the mrp logic that would affect the current mrp plan. These could be changes in inventory levels or part requirements, or engineering changes to the bill of materials, for instance. New mrp requirements are generated only for parts affected by these transactions. Since only those parts needing replanning are affected, the computer maintains the rest of the existing mrp plan that is still valid. This substantially reduces computer processing time to a point where nightly runs become feasible.

Net change mrp can be used to *simulate* master production schedule changes. The output of this simulation run would be a series of reschedule messages that point out any difficulties that the master production schedule change causes. This simulation procedure can be done quickly (as opposed to an entire regenerative run), and can be easily revised or returned to the original plan if its results are not suitable.

The net change requirements planning only goes as far down the product structure or bill of materials levels as it finds necessary to cover all changed requirements. This process is referred to as a partial explosion.

There are many more-detailed differences in data-base design, transaction posting, and reporting methods between regenerative and net change mrp systems. These differences are too technical to be discussed here, but can be found in some of the references at the end of this volume.

As a result of the differences outlined, the effects of each system are that

> The regenerative approach means that information contained in the weekly requirements run is out of date practically as soon as it is published—certainly by the end of the week.
>
> The net change approach allows the requirements plan to be up-to-date on a daily basis.

Regenerative or Net Change mrp Selection Citeria

There are environmental considerations that can influence whether a manufacturing company selects a net change or regenerative approach to mrp. In general, these can be summarized as follows:

A regenerative mrp system is best suited for:

- relatively uncomplicated product designs;
- long and stable production runs;
- few design or engineering changes;
- a stable purchased part supply environment;
- few changes in the manufacturing environment.

A net change mrp system is best suited for:

- complicated product designs;

short production runs of many products;

frequent design and engineering changes;

an unstable purchased part supply environment;

many parts to master schedule.

Either approach to material requirements planning can be used. The net change approach, however, carries a caveat with it:

> *Net change mrp, by reacting to transactions occurring every hour of every day, can sometimes generate new requirements faster than humans can adapt to them. Priority and schedule changes take time to be effected. A net change system can be too "nervous" and have the factory in a constant state of change. Some damping of the system through exception message filters is necessary to allow it to work efficiently.*

I feel that a company should spend 1–2 years installing, working with, and understanding a good regenerative mrp system. Then, perhaps, it will be ready to convert to a net change system and reap the benefits of having more timely information on which to base its manufacturing planning and control efforts. There are many MRP software packages available today that can accommodate this scenario since they can be operated in either a regenerative or net change mode.

Lot Sizing in mrp

Having an mrp system makes it possible to order parts in exactly the quantities needed to conform to the requirements generated by the mrp logic. This is known as lot-for-lot lot sizing.

This may or may not be an attractive feature of mrp depending on the environment in which the system is to be installed. If ordering costs are high, or if ordering *volume* is high, it may be beneficial to examine types of lot sizing that will lower the ordering cost but at the penalty of carrying more inventory.

EOQ lot sizing attempts to balance ordering cost with carrying cost. The EOQ formula is shown for reference below:

$$\text{Lot size} = \sqrt{\frac{2AS}{iC}}$$

where

A = Annual part sales

S = Setup or ordering cost

i = Inventory carrying cost

C = Part unit cost

EOQ is based on the assumption that part demand is at a constant rate, when in fact it seldom is, especially for lower-level parts. I use the words "attempts to balance" because the EOQ formula includes a term for inventory carrying cost. This is a very elusive figure; few firms really *know* their inventory carrying costs with any degree of accuracy. Twenty-five percent is a commonly cited inventory carrying cost percentage, but recent studies point to an inventory carrying cost more in the range of 30–40% as being appropriate in today's current manufacturing and financial environment.

EOQ is not a reasonable lot-sizing model with mrp, since lumpy demand violates one of the basic assumptions of the EOQ model. Various attempts have been made to get around this by attempting to shorten the demand period into 4–6 week periods, for instance, instead of a year or a quarter. Nonetheless, EOQ remains a poor lot-sizing choice for mrp systems.

The lot-sizing techniques applicable to mrp systems for the most part are based on the concept of discrete lot sizing, i.e., ordering exactly what is required to make a certain amount of parts in a given time period. The simplest of the discrete lot-sizing methods is lot-for-lot wherein the exact net material requirements are ordered for each time period. As a result no inventory is carried, unless safety stock is necessary. Fixed order quantity (FOQ) would order the same amount of parts every time, no matter how many are required. Fixed period lot sizing is a variation of this, and it groups order quantities in lot sizes exactly matching requirements for a constant future time period—e.g., for the next 4 weeks. Period order quantity (POQ) attempts to compute the least cost ordering period based on the EOQ concept. Again, this algorithm suffers in the mrp environment due to discrete and lumpy demand.

Several other lot-sizing techniques are usable such as part period balancing, least total cost, least unit cost, and the esoteric Wagner–Whitin algorithm. These dynamic methods all have the disadvantages of being hard for people to understand and of causing excessive system "nervousness," since order quantities are constantly subject to change.

After the part requirements are lot sized by any of these algorithms, they can be altered by an order modifier such as a minimum, maximum, or "multiple." These modifiers may be established for parts based on packaging or part discount (price break) quantities.

The primary benefits from mrp, or 80–90% of the improvement in a company's manufacturing planning and control system, will come from using mrp with any of the simple lot-sizing techniques. Different lot-sizing techniques can be shown to be superior in different situations. For the initial installation and start-up of an mrp system, I generally advocate that lot-for-lot or fixed period lot sizing be used, with the selection of the fixed period length made by several competent, experienced purchasing and manufacturing executives. With the fixed period method, parts might be grouped into families based on ordering and storage/handling costs, and several different time periods used. In any event, selection of the "ultimate" lot-sizing technique is in my view a "stage 3 or 4" improvement for mrp, if an improvement at all. Again, most of the improvement to be gained from a well-run mrp system comes from factors other than the lot-sizing decision—mrp's primary value is that of a scheduling and priority planning tool.

The Role of the Planner

The planner(s) plays a vital role in keeping an MRP system working smoothly and effectively. The planner is responsible for working with the MRP system reports to carry out the following tasks.

Review and release planned mrp orders. Here, the planner reviews the planned order in light of any knowledge he or she might possess regarding unusual conditions in the plant or in Purchasing that would dictate modifying the planned order in any way. If there are no changes necessary, the planned order is released by the planner on the planned release date. This release action changes the planned order to a scheduled receipt and allocates the necessary quantity of parts in the stock room.

Review and react to reschedule messages generated by the master production schedule or mrp logic. These messages in most cases involve rescheduling or cancellation actions that should be tempered with human judgment before they are enacted. If an item is to be rescheduled, the planner should work up the bill of materials using single-level pegging in an attempt to solve the problem. The planner should change the master production schedule only as a last resort.

Work with shop floor control information to review open orders for progress against plan and for possible rescheduling. Here, the planner is the vital communications link between the foremen on the shop floor and the MRP system. He or she is in constant touch with shop floor people to determine if parts can be kept on schedule. If not, the planner must reschedule the past due (late) part, and evaluate whether any other parts affected by this action also must be rescheduled.

Work with shop floor and capacity planners to review planned orders against capacity resources. Here, the planner can peg requirements in a given work center, for instance, and move planned orders by use of the firm planned order to smooth capacity requirements.

In short, the planner is the vital human link in the system who deals with the master scheduler along with shop floor and inventory control people to control at the part number level questions of part coverage and work center capacity.

5

Production Planning and the Master Production Schedule

In the preceding chapters we have examined the basic data and logic required to plan material requirements for any given part. Now this must be related to the overall task of production planning and scheduling for the entire factory. These tasks are accomplished through the production planning process, which leads to the creation of a master production schedule that drives the material requirements planning process (Fig. 5-1).

The Production Planning Process

The production plan is top management's input to the master production schedule. It is a general statement of the desired rate of production for the plant such as cases/day, units/day or week, tons/hour, etc. However, unlike bucketed time-phased information for specific product orders, the production plan is only a rough statement of the rate of plant production. Generally, the production plan is for products that are aggregated into product families or groups of similar parts. In a shoe factory, for instance, a typical production plan might be stated as in Fig. 5-2.

Frequently, fairly accurate marketing forecasts exist for groups or families of products. These forecasts serve as a vital input to the production planning process. As we will see in the next chapter, if these group forecasts are used with planning bills of materials or labor, we can very quickly perform rough cut or resource capacity planning that will tell us in broad terms if our production plan is realistic and achievable.

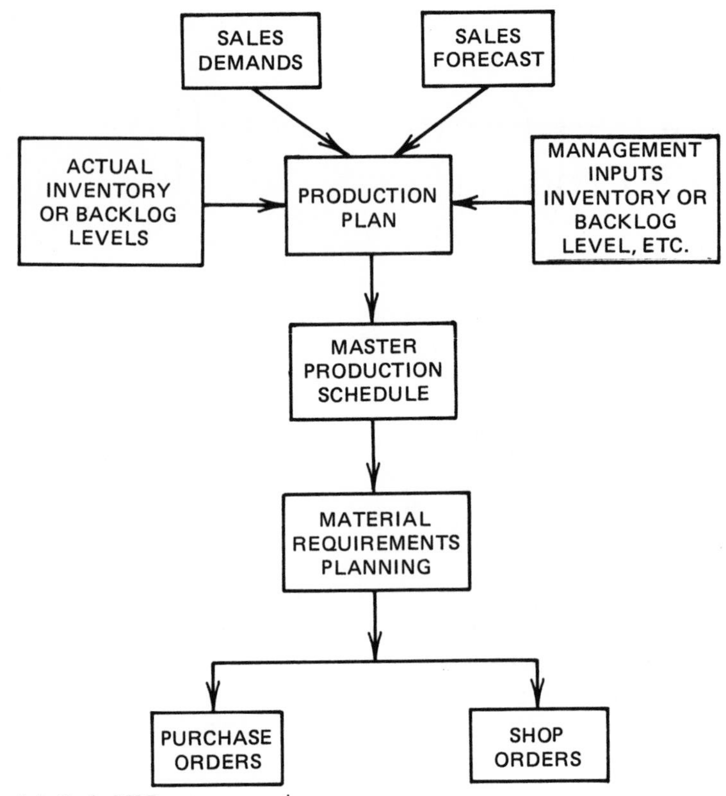

Fig. 5-1. Basic MRP system overview.

	Cases/day
Welts	100
Mocs	75
Cements	120
Sneakers	50

Fig. 5-2. A shoe factory's production plan, 1981.

What the Master Production Schedule Is

The master production schedule is the final statement of planned production by a company. It is the job of the master scheduler to translate the broad family or group rate-based production plan into a time-phased master production schedule calling for a specific quantity of top-level products to be manufactured at a given time.

The master production schedule is the result of an iterative communication and negotiation process that occurs between Marketing, Manufacturing, and Finance or other top management people. It represents a *commitment* by Manufacturing to produce specific items by a certain date. The production planning process used to arrive at the master production schedule is shown in Fig. 5-3.

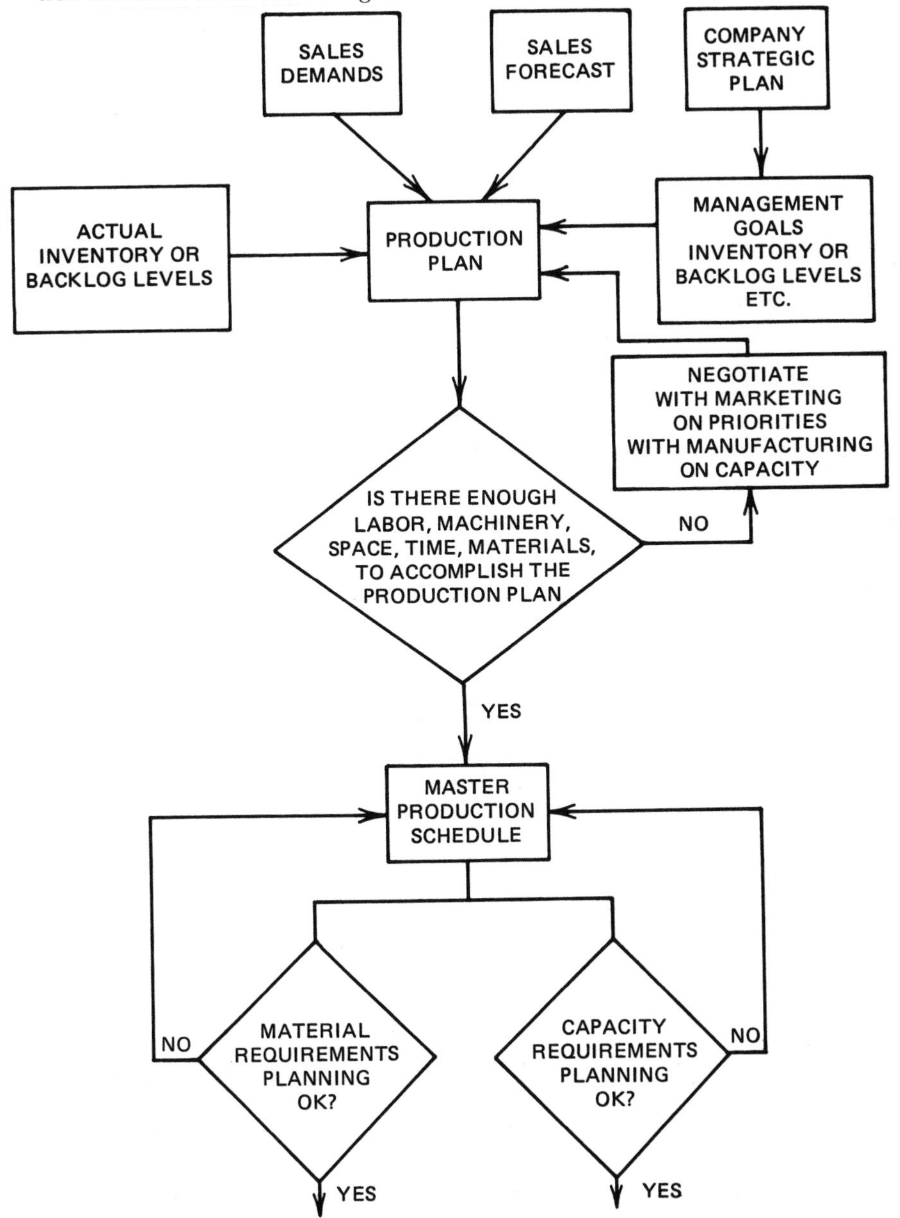

Fig. 5-3. The production planning process.

Note that before the master production schedule is agreed to by Manufacturing, they will have used other modules of the manufacturing planning and control system to see if, even on a rough cut basis, they have the capacity to do the job. The master scheduler also has the responsibility to check that the master production schedule, when aggregated to the product group level, agrees (within defined limits) with the established production plan created initially. The linkages that tie the production plan and master production schedule together are planning bills of materials or labor.

Note also that the master production schedule becomes the vital control center for the company's manufacturing planning and control system (Fig. 5-4). The master production schedule, along with its more macro rate-based production plan as a predecessor, is the interface between Marketing and Manufacturing.

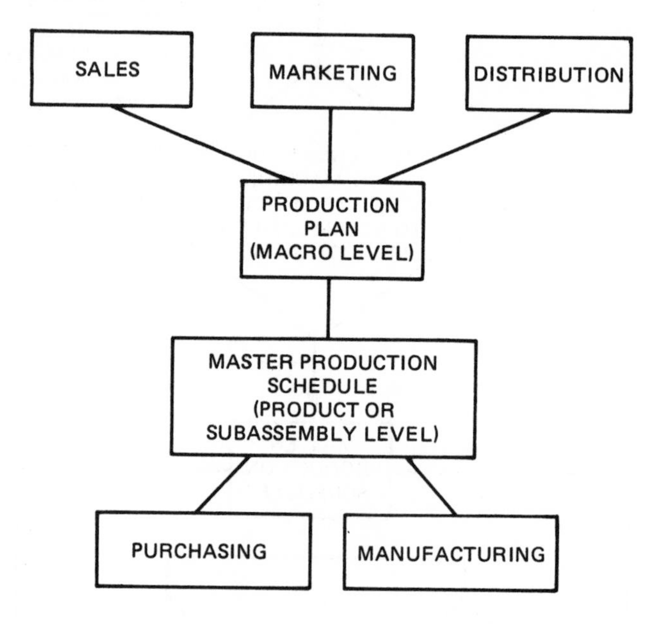

Fig. 5-4. Master production schedule relationships.

The General Master Production Schedule Format

While there can be many different master production schedule formats, depending on considerations we will examine shortly, the basic master production schedule format (Fig. 5-5) is very similar to the mrp format illustrated earlier. The master production schedule shows in the master schedule row that a specific quantity of manufactured items is needed in a certain time period.

	Time period				
	1	2	3	4	5
Forecast (or backlog)					
Gross requirement					
Inventory on hand					
Available to promise					
Master schedule					

Fig. 5-5. Basic master production schedule format.

The Function of the Master Production Schedule

Essentially, the master production schedule report shows the aggregation of gross requirements from all demand sources. Then, a netting and lead time offsetting process similar to that of the mrp logic is used to arrive at net manufacturing requirements, the output of the master production schedule.

The inputs to the master production schedule in the form of orders and forecasts are demands on the company's resources. Depending on top management's stocking policy and inventory or backlog goals, customer demand may not directly translate to demand on manufacturing.

Many companies try to live without a master production schedule by dumping gross sales requirements into the mrp logic, or else by using planners to perform the netting of the gross sales requirements before passing them on to the mrp logic. While the latter method might be correct, it is far too cumbersome a task to perform manually for more than a few products per planner. Since the logic is simple to program, why not let the computer do it? The former procedure, however, is not correct. Putting gross requirements into the mrp logic will ultimately result in a mismatch of priorities and requirements between Manufacturing and Sales. For instance, there might be gross requirements on a part for 10 pieces. If there are 8 in inventory or 8 due in shortly as scheduled receipts from Manufacturing, true demand on Manufacturing should be 2, not the 10 that would result from dumping the gross requirements into mrp.

There has been a great deal of progress made recently in the area of master scheduling. Now, we are starting to see multilevel master production schedules where the final product (Level 0) is scheduled for assembly in the first or top-level master production schedule, then the second-level subassemblies (Level 1) or modules are also master scheduled in the second master schedule by an mrp-like netting process. The output of this second Level-1 master production schedule is then fed into mrp for material planning and shop labor and machine loading. The demand for the company's products and its manufacturing environment determine how the company's products are to be master scheduled.

Types of Product Demand

The type of product demand each company faces plays a large part in how the company creates its master production schedule. We will now look at possible sources of manufacturing demand and how they influence the master production schedule logic and format.

Demand for manufactured products can originate from any number of sources—both internal and external. Under *external* sources, i.e., those originating outside the company, we find the following:

- Direct (retail) customer orders
- Distributor/jobber orders
- Company-owned distribution center orders

Internally, i.e., from within the company, demand can originate from:

- Sales forecasts from marketing
- Finished goods stock leveling or seasonal inventory building or finished goods safety stock requirements or finished goods inventory replenishment
- Orders from other company-owned plants

Note that this demand can be for end items or replacement service parts.

The Manufacturing Environment

There are several ways to classify the manufacturing environment in which the company exists. The *major* question is does the company *make-to-stock*—that is, carry a finished goods inventory—or does it *make-to-order.* This classification can be further broken down in the form of a 2 ×

2 matrix by showing whether the company builds a few products or many products. The above picture then looks like the matrix in Fig. 5-6.

Manufacturing
End Items or Replacement Parts

	Make to stock	Make to order
Few products	a	b
Many products	c	d

Fig. 5-6. The manufacturing environment.

Naturally, things are seldom as clear-cut within a company as the matrix in Fig. 5-6 illustrates. In fact, most companies are a mixture of a make-to-stock and make-to-order environment. However, the company's dominant manufacturing mode must be the prime influence on the way the master production schedule is organized, i.e., what is master scheduled, and how the master production schedule logic operates, and how the master production schedule report is formatted.

Note that in a make-to-stock situation future demand is usually based on sales history, with manufacturing demand being the result of netting total forecast demand against the existing finished goods inventory. Note also that some companies, when faced with highly seasonal finished goods demand, make-to-stock during slack sales months to level production during the year.

In a make-to-order situation, future demand is usually in hand as order backlog, which may stretch out months or years into the future. Typically, no finished goods inventory exists in a make-to-order environment. There is no question of what to manufacture here! Question usually only focus on the *priority* of each order in the backlog.

If your company is a make-to-order company, is there a strategy to deal with the possibility of the backlog drying up? How will your company deal with this: make-to-stock, or cease or cut production? What are the trade offs for your manufacturing organization, and in your industry?

What to Master Schedule

The type of demand a company faces, and, indeed, the type of industry the company is in, influences the level at which products are master scheduled and the format of the master production schedule report. In addition, the type of product the company produces also influences what should be master scheduled. Examples of common situations follow.

Make-to-Stock

Here, the company is producing to satisfy inventory level goals set by top management based on forecasted sales expectations and desired customer service levels.

Make-to-Stock—A "Few" Products. For companies that produce a "few" such products—less than about 500—the end items are usually master scheduled. Any seasonality in the forecast or required inventory levels should be leveled across the year before the manufacturing requirements are input to the master production schedule.

Make-to-Stock—Many Products. When more than about 500 end items exist, it becomes too impractical to master schedule all this detail. In this case, a popular technique is to aggregate products into lines, or product classes, or families. The aggregated product class requirements become the items master scheduled. Then, to assign end item quantities to be scheduled for manufacture by the mrp module, end items are usually scheduled based on their historical percentage of product class sales.

Where Many Product Options Exist. The same technique as in the preceding paragraph is commonly used where many end items can be made from different combinations of optional subassembly modules. Here the possible number of Level-0 items is again too large to effectively master schedule. Instead, the Level-1 subassemblies or basic modules are master scheduled, based on their historical percentage of usage on end items.

Make-to Order

In a make-to-order environment, Manufacturing does not usually schedule a product for production until Sales has the customer's order in hand. More specifically, Manufacturing needs the specification from the customer order as to product options, design features, etc. Thus the customer order backlog becomes management's input to the master production schedule. Note that in a make-to-order environment, the "forecast" or backlog is always known with certainty—not subject to error, only cancellation.

A policy for manufacturing strategy must be established in make-to-order companies regarding what happens if the backlog is not sufficient to cover the total manufacturing lead time of each product. If it is not, will manufacturing activity levels—throughput capacity—be tailored to the backlog level? Or will production be established at a constant level, and will any gap between backlog and production rate be filled with in-house-generated orders for forecasted (most commonly built) "spec" orders? The answer to this key question depends on the volatility of each industry's sales and the predictability of design specifications, which in turn are influenced by the rate of technological change and competition within each industry.

Differentiating between "few" and "many" can be performed in the same manner previously outlined for the make-to-stock situation, that is, using the techniques of master scheduling Level-1 modules through product family grouping and analysis.

The Master Production Schedule Planning Horizon

Most master production schedules extend over a 52-week, 1-year period, with perhaps a second year shown in months or quarters. Intuitively, the master production schedule planning horizon must extend out far enough to equal or to exceed the cumulative or "stacked" lead times of raw material procurement lead time and/or manufacturing lead time. This stacked lead time defines the end item's critical time fence. The *critical time fence* is the time point for a product inside which changes to the master production schedule become increasingly prohibitive because work already started must be stopped or changed in some manner. The closer a product is to being completed, the more value is added to it, and the more expensive it becomes to change the manufacturing order in any way.

Cumulative lead time is the total time it takes to completely produce a part, including the procurement of raw materials and parts. It can be found by analysis of lead times required to make or procure each part in an assembly.

If the product structure for an assembly is tipped on its side and lead times are attached to each part or subassembly, the cumulative or stacked lead time will be the route through the branches that adds up to the highest total (Fig. 5-1). Here, the cumulative lead time for assembly H is 5 weeks. This time consists of the time spent in

Procuring part A	2 weeks
Making subassembly G	2 weeks
Making assembly H	1 week
H's cumulative lead time	5 weeks

Note that the 5-week path shown in Fig. 5-7 is the "critical path" for this assembly. This is a good example of the fact that mrp is a scheduling tool and is based on the same general principles as the critical path method (CPM) of scheduling. Indeed, Fig. 5-7 is similar to the familiar CPM diagram.

The critical path is the chain of events that has the longest sequence and, therefore, takes the most time to accomplish in order to meet the end objective. Any events not on this path have some slack time available in which to be completed before a "parallel" event on the critical path is completed. In Fig. 5-7, parts C and D each have 1 week of slack time in relationship to part A. By *backscheduling* from the due date, mrp always puts the slack time ahead of each event—delaying the use of labor and materials until the latest possible moment—thus minimizing inventory investment.

The relationship of cumulative lead time to master production schedule time zones and the original time fence is shown in Fig. 5-8. The critical time fence defines the point at which master production schedule changes can no longer be made easily. If we tried to change the master schedule for part H in week 4, it would be too late to alter the fact that the part A's necessary to make H were ordered 1 week before.

While the master production schedule inside the critical time fence can be changed, changes in this time zone should only be made after careful consideration is given to customer priority, cost trade offs, and the ability of the people on the shop floor or in Purchasing to accommodate such changes. Generally, changes to manufacturing plans can be made outside the critical time fence with no penalty.

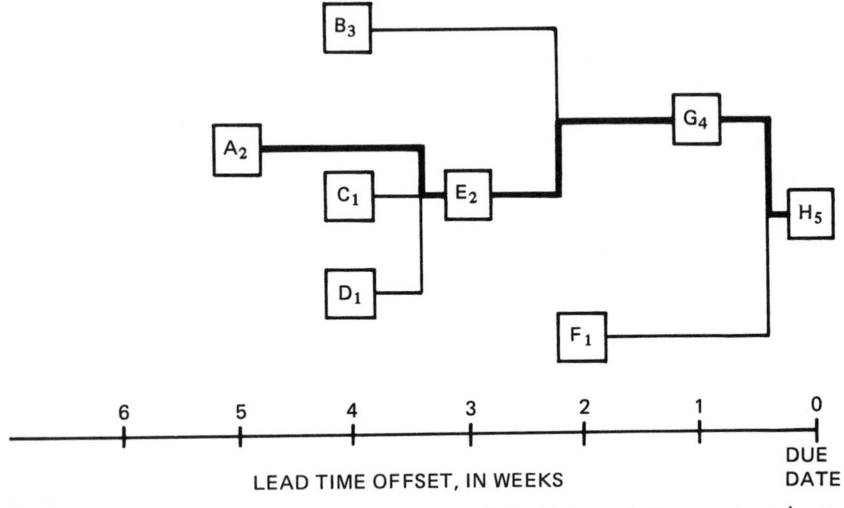

Fig. 5-7. Cumulative or stacked lead time for assembly H. (A through H are part numbers; the number in the box is the weeks of lead time required per part number. Bold line is critical path or cumulative lead time.)

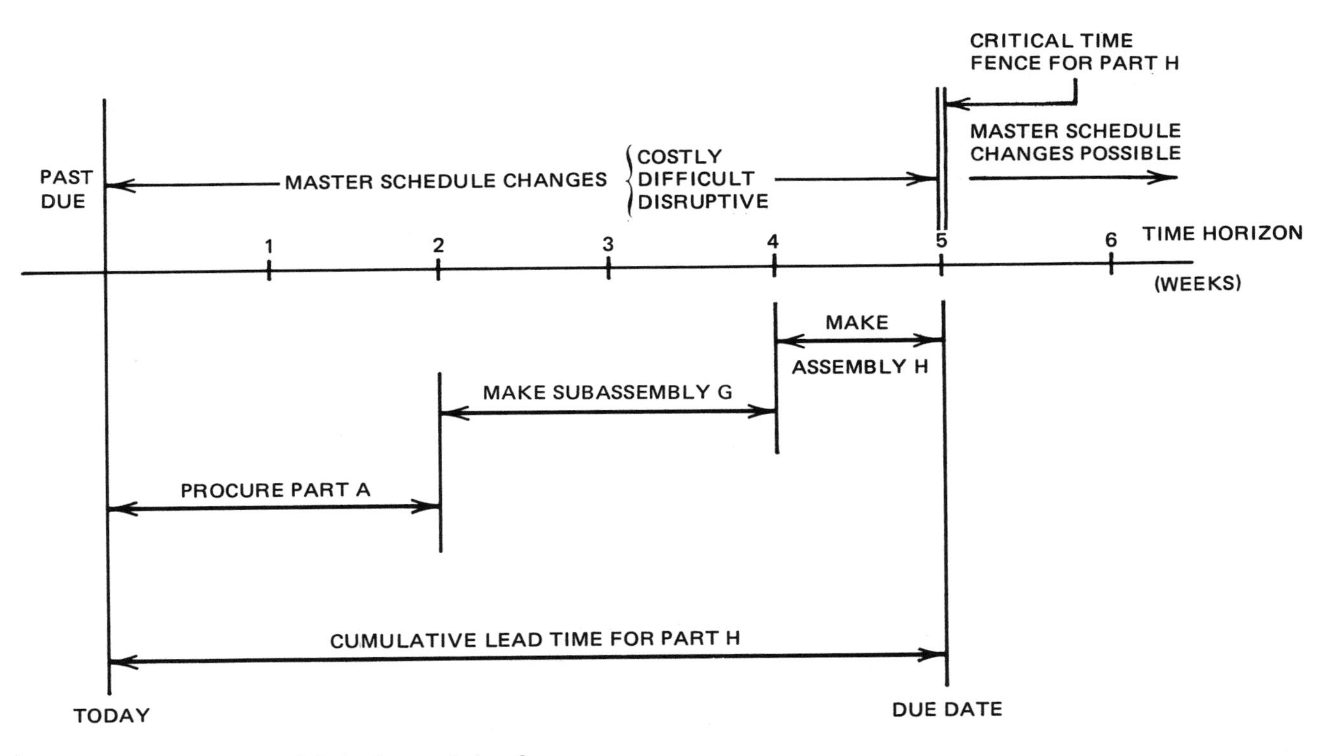

Fig. 5-8. Master production schedule horizon and time fence.

Some Master Production Schedule Formats and Definitions

We will first give definitions of some terms commonly included in master production schedule reports.

The purpose of the master production schedule report is to show the planner the relevant data that influence the calculation of the net amount of product to be manufactured.

Demand figures:

Sales forecast

Direct customer orders

Distribution center orders

As we have seen, demands on the company's manufacturing resources are input to the master production schedule. These are self-explanatory and can be shown in detail by line item before they are aggregated by the master production schedule logic to show total demand on the company. Usually, the demand is shown against the forecast of the part that was developed by Marketing. An example of this is shown in Fig. 5-9.

	Week		
	1	2	3
Sales forecast	50	60	65
Direct customer orders	60	50	40
Distribution center orders	20	15	20
Total demand	80	65	60

Fig. 5-9. Master production schedule demand aggregation.

Projected figures:

Projected available balance

Projected backlog

Available to promise

Projected Available Balance

In a make-to-stock situation, the projected available balance is usually shown as a master production schedule report line item. This is the standard inventory calculation where for each time period

Beginning available balance
+ Scheduled receipts
− Shipments

= Projected available balance

Note that this calculation starts with the on-hand inventory balance known for the current time period. An example of this calculation is shown in Fig. 5-10.

	On-hand	Week 1	2	3	4
Scheduled receipts		30	0	40	45
Scheduled shipments		20	25	30	30
Projected available balance	30	40	15	25	40

Fig. 5-10. Master production schedule projected available balance calculation.

Projected Backlog

In the make-to-order situation, projected backlog is usually shown on the master production schedule. Its logic is in the same pattern as the projected available balance in that for each time period

Beginning backlog
+ Booked orders
− Shipments

= Projected backlog

An example is shown in Fig. 5-11.

	Current Backlog	Week 1	2	3	4
Booked orders		80	70	50	30
Schedule shipments		70	60	50	40
Projected backlog	100	110	120	120	110

Fig. 5-11. Master production schedule projected backlog calculation.

Available to Promise Figure

This figure is valuable to Marketing or order-entry people in assigning products or delivery dates to customers in make-to-order situations. This calculation simply nets out the total projected manufacturing schedule against products spoken for by customers in order to show how many products in a given time period are actually still available to promise to a customer, as shown below:

Beginning available inventory
+ Scheduled receipts
− Customer orders

= Available to promise

An example is given in Fig. 5-12.

		Week			
	On-hand	1	2	3	4
Customer orders		90	60	40	30
Scheduled receipts or master schedule		50	50	50	50
Available to promise	60	20	10	20	40

Fig. 5-12. Master production schedule available to promise calculation.

Additional master production schedule line item definitions are:

Scheduled Receipt: an incoming order for parts previously ordered or master scheduled and either in transit or work in process on the shop floor.

Master Schedule: a production order to produce parts, subassemblies, or assemblies. The output of the master production schedule netting and time-phasing process.

Receipt Dates: a receipt date indicates the time the product is available for use or shipment.

Start Date: the start date is the date the order is to be released to a vendor or the shop floor.

It is important to realize that there is no standard master production schedule format. Indeed, within the majority of companies that are both make-to-stock and make-to-order, it is reasonable to have more than one master production schedule format, depending on the type of product being made. What is most important, though, is to have a set of agreed upon master production schedule line item definitions, and for everyone concerned with the master production schedule to know how the master production schedule netting process works and how the calculated master production schedule line items are derived.

The Role of the Master Scheduler

The master scheduler plays a *key* role in the proper operation of any manufacturing planning and control system. He or she is the controller of the entire operation. While he or she may not have the command authority that top management has, the master scheduler's role is to provide a timely and loud warning that the production plan and master production schedule are either out of balance or that the company's production schedule is in trouble due to a shortage of labor, material, capacity, or time.

For instance, if Marketing tries to overload the master production schedule in the production planning process, or recent manufacturing shortfalls have resulted in a large amount of past-due work, the master scheduler must voice a firm "No" to Marketing's demands for more production (assuming the plant is running at maximum capacity).

Rather, he or she need not say no, but not *now*. Not all at once. The master scheduler must go back to Marketing and call for a rescheduling of order priorities. All that he or she can say to Marketing is: "You cannot have everything you originally wanted in the time frame you wanted it. Therefore, which orders or products do you want first?" Or, in another situation, he or she might have to say: "Since we are at maximum capacity, existing orders must be scheduled out to accommodate the past-due work we must make up." Please note again, the master scheduler does not and should not have decision-making power that belongs to top management and/or to Marketing. He or she is, however, the company's warning system for the inevitable daily problems that arise in manufacturing.

Master Production Schedule Management Techniques

As a tool to aid the master scheduler in managing the master production schedule, the master production schedule netting logic should generate exception messages that alert the planner or master scheduler as to what action to take. These rescheduling exception messages can be triggered by management-set limits that react to such line items as the projected available balance, or to master schedule order start dates that release an order that is already past due because it was started too late, i.e., in less than the cumulative lead time for the product.

It is important that filters be established to limit reschedule messages to those that really count. A planner would probably not want a reschedule message that told him to reschedule-in 2 days a product with an 18 week cumulative lead time. He or she *would* want to see the message if the reschedule-in time for this part was 2 weeks!

Master production schedule reschedule-exception messages generally tell the master scheduler one of three main things:

1 / Move (reschedule) the planned order in.

2 / Move (reschedule) the planned order out.

3 / Cancel the planned order; it's not needed at all.

Each master production schedule action taken in response to a rescheduling exception message has an effect on the manufacturing picture.

1 / Moving a due date on a *released* order in to where the manufacturing time is less than the cumulative lead time compresses the lead time somewhere in the material procurement or manufacturing process. This has the same effect as planning an order to start "past due" (late) where the time available to produce the part is less than the cumulative manufacturing lead time. This tactic often can be made to work in isolated cases if the order is expedited, since *average* lead times are used for scheduling in the master production schedule/mrp systems. The danger here, however, is that this remedy will become abused, and the shop floor then regresses back to the old informal system where *every order is top priority.*

2 / Pushing the due date out on a *released* order means that we could have excess raw material and parts inventory, or partially completed units on the shop floor for longer than necessary. This is the lesser evil, but, of course, negates the reason for a good manufacturing planning and control system—the reduction of all inventories.

The important thing is to keep master production schedule changes to a *minimum*—preferably none within the critical time fence. Each master production schedule change generates a series of action messages to the shop floor, Purchasing, Inventory Control, stock rooms, etc. If master production schedule changes abound, action messages will increase geometrically until people and the system become congested.

A useful management technique is that of overstating the master production schedule. Overplanning or overstating the master production schedule is really allowing for some safety stock at whatever level items are master scheduled. This is a valid procedure, but note that excess inventory is created with this technique. In effect, vital capacity may be used to overproduce one product at the possible expense of *not* being able to produce a more- or equally needed product. However, unlike safety stocking at the individual part level, overstating the master production schedule means that there will exist lots of evenly matched end item sets, subassembly sets, or module sets, whatever is being master scheduled. Thus, part orders or inventories do not get unbalanced in relation to bill of materials structures and top-level part requirements.

Keeping the Master Production Schedule Viable

The most important task of the master scheduler is to keep the master production schedule realistic and, therefore, viable. The task here centers around four essential items:

1 / Maintaining accurate master production schedule data input.

2 / Proper treatment of missed or "past-due" production.

3 / Proper capacity planning.

4 / Proper shop floor feedback and tracking.

To some extent these are all related, as can be seen by examining each separately.

1 / The data input to the master production schedule and mrp must be kept accurate. This means utilizing accurate lead times, inventory balance figures, due dates, shop calendars, product class families, bill of materials, etc. Rigorous pursuit of errors and "zero defects" should be the goal for all people who control data input to the manufacturing planning and control system. Errors creep into systems slowly and insidiously until they are so rampant that the system's output becomes meaningless. A "dirty" data base can take months to clean up—a thankless task. It is important to have good control at data-entry points with methods such as check digits and operator access codes, methods that allow only certain levels of management and/or people at certain terminals to make important changes, and maintenance reports and clear audit trails that allow reconciliation of problems. These steps complicate the system software and may slow down data entry, but generally pay off in a much higher level of data accuracy than is found in systems that do not contain such safeguards.

2 / With regard to "past-due" production (or missed production, or "slips"), management must bite the bullet and adopt a firm policy concerning rescheduling. Simply stated, capacity is *perishable.* Once you have gone past today, you cannot produce today's lost production in anything but *tomorrow.* Yesterday has no production capacity. The task that must be performed by the master scheduler is to add the uncompleted or past-due requirements to the present time period's requirements. At the same time (assuming maximum capacity), the master scheduler must reschedule future requirements by rolling forward previously scheduled work until the past due can be worked off. To illustrate (see Fig. 5-13), if we are in period 1, with 10 units of past-due or missed production, and the next four time periods are laid out as indicated in the figure, then the 10 past-due units must be moved into week 1 and 10 units moved out through all 4 weeks and into period 5 as shown. We normally assume the 10 past-due production units have a higher priority than those scheduled in periods 1–5. But this might *not* be the case, and a *priority check* with Marketing might indicate those 10 past-due units could be put aside and finished in week 5. If this were the case, the original master production schedule would still be correct for weeks 1–4.

Getting top management to acknowledge the past-due production capacity problem and the need for rescheduling lost production is one of the toughest problems manufacturing executives face. However, until management faces this problem head on, the master production schedule will never be valid and achievable.

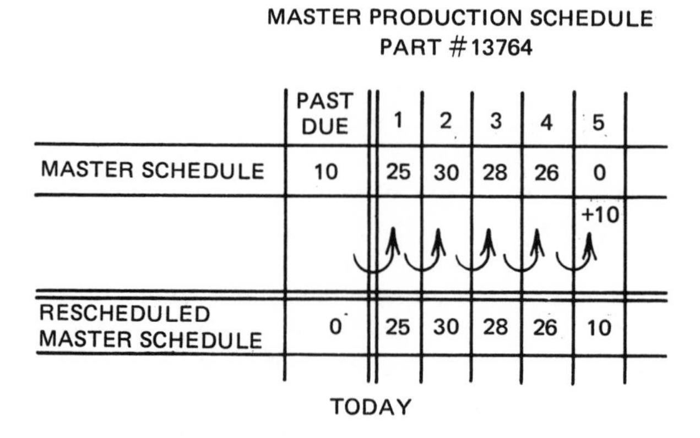

Fig. 5-13. Rescheduling past-due work. The number in each bucket is the net manufacturing quantity to be produced.

3 / Concurrent with the preceding message, the master production schedule can never be allowed to become management's "wish list"—so

unrealistically overstated that it is impossible to achieve. The key to maintaining control over the master production schedule is to plan capacity properly. Armed with accurate output from the production planning process (resource requirements planning) and from capacity requirements planning, Manufacturing must be able to tell top management the job cannot be done with the present resources, if that is indeed the case.

6

Capacity Planning

Capacity planning is the act of balancing the company's manufacturing resources against the demand for its products. These resources consist of land, manufacturing equipment, labor manufacturing space, and whatever materials are required for production. Capacity planning can be approached at three levels, as illustrated in Fig. 6-1. At *each* of these levels, capacity planning seeks to balance manufacturing *resources* against manufacturing *requirements.*

Level of Detail	Type of Capacity Planning	Time Horizon
Low, long range	Resource requirements (rough cut) planning (RRP)	1–5 years
Medium range	Capacity requirements planning (CRP)	1–18 months
High, short range	Operational sequencing planning (OSP)	1 day to 4 weeks

Fig. 6-1. An overview of capacity planning.

Resource Requirements Planning

Resource requirements planning (RRP) for manufacturing really should be linked to the company's strategic planning. Most company's master schedules do not cover 5 years out on the time horizon. Instead, most companies carry a weekly bucketed master production schedule for 1 year. Beyond the first year, it is useful to show aggregated figures for larger time buckets. Two possible scenarios are shown in Fig. 6-2.

Planning Horizon

	Year				
	1	2	3	4	5
#1	Weekly	Monthly	Monthly	Quarterly	Quarterly
#2	Weekly	Quarterly	Quarterly	Yearly	Yearly

Fig. 6-2. Possible master schedule horizons for capacity planning.

RRP is a "macro" or "rough cut" examination of manufacturing resource requirements. It is based on the same general logic as mrp but uses more averaged data for material and labor or other resource planning. The result of RRP is a computer report that provides approximate answers to such manufacturing questions about the next 5 years as:

> Is there enough floor space for inventory or manufacturing equipment?
>
> Is there enough machine capacity?
>
> Is there enough manpower in the right crafts and at the needed skill levels?
>
> Can we manufacture what our company's 5-year marketing plan requires? If not, what, when, and how do we need to change?

The key to this macro-level planning is to use *planning* bills of labor or materials as surrogates for part-specific routing or bills of materials in the time-phased requirements explosions. These planning bills will be discussed after the following background material.

Usually the 5-year sales forecast will consist of aggregated numbers by product class or family over the next six to eight quarters, and maybe only a total year sales estimate for years four and five. These sales estimates should be in *units,* or be convertible to units, based on some average per unit dollar value.

With this sales forecast as a base, through careful examination of several bills of materials in each product class and an averaging process, a planning bill for whatever resource you are examining can be created for each of the aggregated sales or product line groups. These planning bills can be as detailed as you need them in order to answer particular questions indigenous to your products, company, or industry. The point to keep in mind here, though, is that RRP is intended to be a *quick* and *simple* process easily subject to simulation runs that answers "what if?" scenario questions.

Often, a product sales group will contain many sales options or will be the aggregation of several different sizes or models within the group. The essence of a *planning* bill is that it is an *average* bill representing the entire product sales group. For rough cut capacity planning, the quantity used for each option or model can be a percentage of the total option offering. An example of a planning bill of materials for a product group is shown in Fig. 6-3.

17″ Bicycle MX Group

Basic 17″ bicycle			Rear view	Seat		Handlebars	
Red	Yellow	Black	mirror	Standard	Racing	Standard	Racing
20%	40%	40%	30%	60%	40%	30%	70%

Fig. 6-3. A planning bill of materials.

In the figure note that the planning percentages are approximate and that for options or groups where one must select either–or, the weighting percentages must sum to 100%. Where there is only an option, only the yes percentage, which will be less than 100%, is shown. In other words, if in the next year we were going to build 100 17″ MX bicycles, the planning bill of materials explosion would result in rough cut material (or part) requirements of

20 Red 17″ basic bicycles
40 Yellow 17″ basic bicycles
40 Black 17″ basic bicycles
30 Rear view mirrors
60 Standard seats
40 Racing seats
30 Standard handlebars
70 Racing handlebars

A planning bill of labor is shown in Fig. 6-4 for a typical family of products. A key assumption in using planning bills is that the existing product family *mix* will not change over the 5-year period. If new products are being planned that are radically different from the company's existing products or that will use a new method of manufacture, then these new products will require their own new estimated planning bills.

RRP is performed without regard to the present work load or materials in the factory, since the time period involved is usually far greater than the time necessary to finish manufacturing the current work in process.

Product Family XYZ

Work center number	Work center	Standard hours per 100 units
100	Grinding	25
200	Drilling	20
300	Milling	50
400	Painting	20
500	Turning	10
600	Welding	5
700	Assembly	10

Fig. 6-4. A planning bill of labor.

Capacity Requirements Planning

Here, we are looking at capacity requirements during the period covered by the master production schedule. The goal of this planning is more likely to be one of leveling production throughout the year rather than considering major changes in plant capacity. In the 0–1-year time range, capacity in most types of industries is reasonably fixed. This is especially so for the first six months, due to the time required for hiring and training, machinery procurement, and plant remodeling. Furthermore, most company's operating budgets are reasonably fixed for the year ahead, and managers usually will not violate these budgets to make unplanned capacity changes.

Even when subcontracting to relieve capacity problems, it takes time to find reliable vendors. While subcontracting may save your company's machine or labor capacity, the entire subcontracting process can *add* extra processing time to the total work in process time for the subcontracted product.

Since the 52-week work time period is usually master scheduled, it is easy to explode these requirements through routings to arrive at the year's "exact" labor and machinery requirements. These requirements can also be aggregated into months or quarters if necessary. Naturally, the master production schedule will change over the year. Barring significant product mix changes that may not be reflected on the master production schedule, however, the capacity requirements plan will be sufficient to obtain a clear picture of the year's manufacturing capacity requirements, to decide whether load leveling is needed, and to determine what key work centers in the manufacturing process have an excess or shortage of capacity. Note that capacity requirements planning takes into account the work load presently in the factory due to released orders, in addition to work contained in planned orders.

What a typical capacity requirements planning report might show if aggregated into weeks is presented in Fig. 6-5. The problem of insufficient capacity in future time periods may be alleviated by shifting the manufacturing capacity requirements through master production

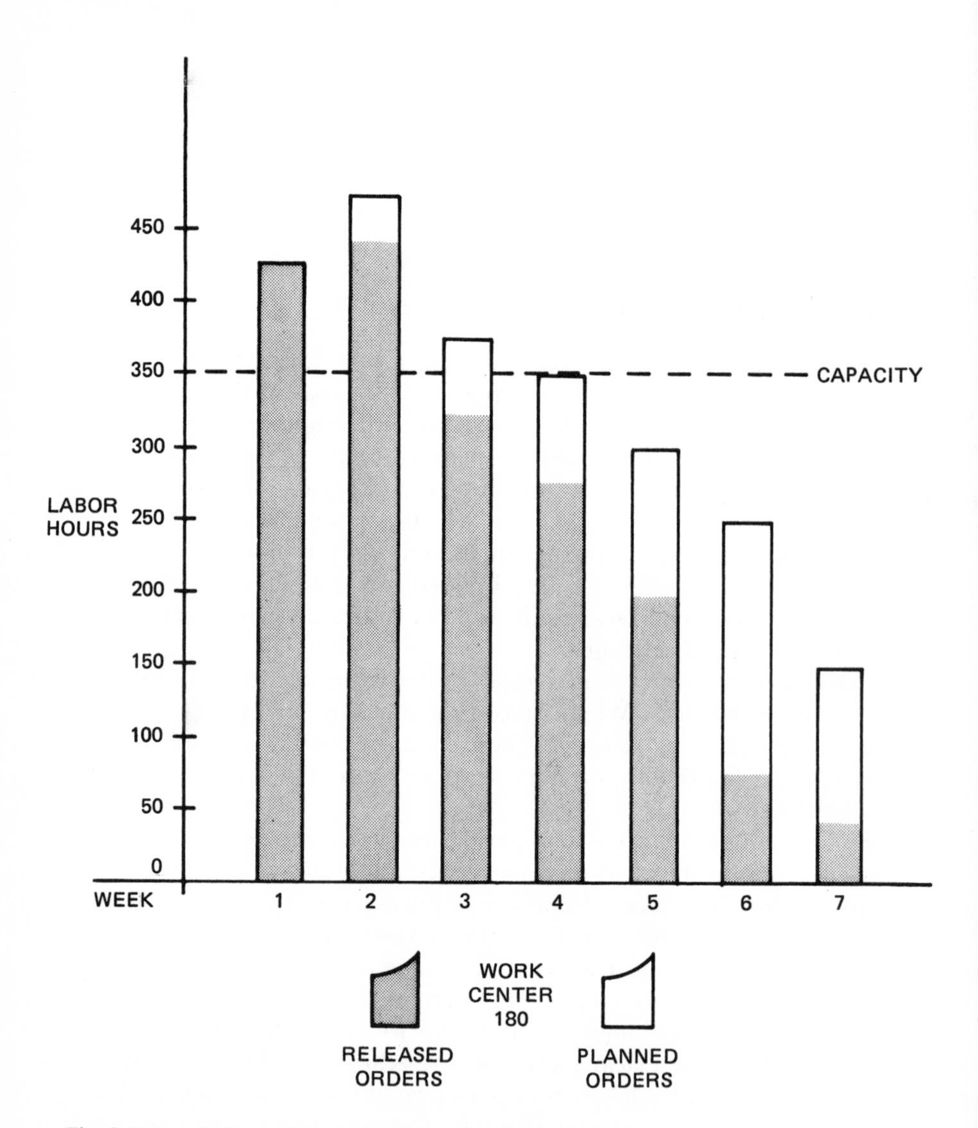

Fig. 6-5. A typical capacity requirements planning report.

schedule changes. Or, as we will see in Chapter 7, it may be possible to alleviate this capacity problem by judicious shifting of lower-level planned orders through the intelligent use of firm planned orders. On the other hand, it might be necessary to establish plans to gain short-term manufacturing capacity through the use of overtime, extra shifts, increased space, subcontracting, adding additional machinery or labor, etc.

In examining capacity based on planned or allowed hours, it is important to know your company's labor efficiency, machine utilization, and other causes of lost time. Labor efficiency is used to measure a worker's performance against the standard labor times shown on a routing. It is defined as

$$\text{Labor efficiency} = \frac{\text{Allowed or planned hours for a task}}{\text{Actual hours taken for a task}}$$

Some limits can be set on this efficiency figure so that job standards can be examined for fairness and/or accuracy if the efficiency on one operation is constantly under 75% or over 130%, for instance.

Machine utilization is defined as

$$\text{Machine utilization} = \frac{\text{Machine usage hours}}{\text{Total available production hours}}$$

Production machinery, like people, are assets that must be managed properly. A good preventive maintenance program will go a long way toward increasing machine availability and, therefore, machine utilization. Time should be allowed both for preventive maintenance and for the unscheduled breakdowns that will always occur no matter how conscientious the maintenance effort. There must also be a conscious effort to achieve high machine utilization so as to provide a quick payback and a high return on investment for the machine.

Obviously, for work on production machinery, a worker cannot put in more productive labor time than the machine is "up" or available for use. Thus machine utilization can be less than available machine time, but both are measured in relation to the total production hours available per day or shift.

There are many causes of lost time or efficiency in a manufacturing operation. Examples of such variable factors are:

training of new personnel
scrap and rework
absenteeism
tardiness
nonproductive time such as
  filling out paper work
  entering data on a CRT or terminal
  clean up

Note that we have factored worker breaktime—a constant figure in most plants—into the planning by using the actual hours available per shift per day, i.e., 7.5 hours instead of 8 hours, in the mrp backscheduling algorithm. In any case, when examining planned capacity requirements for manpower planning, one should be aware of how these lost time factors can affect the labor input required. For instance, assume the following representative labor loss figures:

Training loss	2%
Scrap and rework	3%
Absenteeism	3%
Nonproductive time	2%
Total lost time	10%

Considering the above losses, if the *planned* hours requirement is 1000 hours per week, then you must *apply* 1100 hours per week to achieve this goal. Thus, instead of 25 men on regular time per week, you will need 28!

One can also consider labor efficiency in manning a work center, or creating worker incentive or reward programs. If a worker is turning out 42 planned hours of work in 37.5 hours per week, and the job standards are accurate and fair, then that worker is performing at efficiency of 112% and his or her pay or promotion opportunities should reflect this above-normal efficiency. One should not plan on *always* obtaining this efficiency from a worker, however, due to the varying job mix he or she may have to work on in a given time period, and also to the natural variability of human beings in their work effort.

The point of capacity requirements planning is that you can establish *now* that you will have a capacity problem in the future, and you can plan *now* to make the necessary changes to your manufacturing resources. Perhaps if sufficient manufacturing capacity cannot be added fast enough, it will be advisable to make plans to temporarily reduce or change the company's marketing, sales, or advertising efforts.

Long-Range Materials Planning

Naturally, material requirements for the year-long master scheduled period are planned in detail with every mrp run. It is important for management to obtain an overview of material requirements:

"exactly" for the coming year in material requirements planning,

on a more macro basis, if needed, over the extended time horizon used in resource requirements planning.

The materials picture that results may focus consideration on important material design changes or substitutions that should be made in future periods as domestic and world materials supply availability and prices change.

Vendor Capacity Planning

Most companies have a critical dependence on vendors for their raw materials. Yet, the subject of *vendor capacity* (all vendors or just one vendor) is usually ignored in materials planning until it is expedite time. Your vendor's manufacturing or processing capacity is just as important to your company as its own capacity.

By looking at your company's future reqirements for purchased items, you can establish whether your present vendors might be able to supply the required goods, whether you should seek out more sources, or whether you should reexamine some make/buy decisions made in the past. Just as important, your vendor(s) will be overjoyed if you will share your company's future procurement requirements with them so they can improve their own capacity and inventory investment decisions, and thus be able to do a better job of supplying your company in the future. Indeed, smart companies make vendors an integral part of their manufacturing planning and control system, delivering time-phased material (capacity) requirements to them on a routine basis. In turn, these smart companies are usually rewarded with more timely deliveries at a lower price!

Short-Term Operational Sequencing/Balancing

Here, we will explore the area of operations scheduling on a daily basis, sometimes down to the foreman level or hourly details. Using the concepts outlined earlier, machine center and work station requirements concerning the number of people and machines required, and even the craft and skill levels of workers, are generated by exploding product requirements through routings. A typical work center operational schedule, or dispatch list as it is more properly known, is shown in Fig. 6-6.

In operational sequencing, we deal with issues such as the need for overtime, machine utilization, the leveling of loads at bottleneck machines or work stations, and the day-to-day problems that always occur when dealing with people and machinery. Previously, in capacity requirements planning and resource requirements planning, it was relatively straightforward to manipulate numbers. Short-term capacity planning to a foreman usually implies one thing: *results* are what count—product out the *door*—not plans. This is a crucial difference, and it demands a different set of task skills, background, and people skills to accomplish. The pressure to make targets is high, the resources limited or fixed for all practical purposes, and the daily problems are unpredictable.

Many companies have had trouble trying to make short-term sequencing work on a day-to-day or week-to-week basis. This is primarily based on the fact that, of necessity, lead time and work in process times are *averages.* The short-term real world deals, however, with specific work arrival times, the queue at the work station 10 AM, the fact that Joe is out this week, two new workers are on the next work station, and the materials are half an hour late. However, good capacity planning that uses realistic standards, along with the proper shop floor feedback and controls, can do a lot to mitigate the day-to-day pressures on the people out on the shop floor.

Finite and Infinite Loading

There are two basic methods of capacity planning that can be used in the short term—infinite and finite loading.

In *infinite* loading, we simply explode the bills of materials and routings—backing up by the various operation times (backward scheduling) to arrive at machine and labor requirements by work center. Work center requirements are reported—whatever they may total—with no regard to what happens if a work center is loaded beyond capacity (Fig. 6-7). That is, the computer logic assumes the plant and all work centers have infinite capacity. By comparing the requirements with known capacities, the computer points out where capacity is short. Then production management must intervene to shift work center loads so that any capacity shortfalls are eliminated.

DAILY DISPATCH LIST

WORK CENTER: 125 DATE: 01/22/81

SHOP ORDER #	PART #	OPERATION #	ORDER QUANTITY	OPERATION		ORDER DUE DATE	HOURS			NEXT WC
				START DATE	DUE DATE		SET UP	EXT. RUN	TOTAL	
0173	4350	20	35	01-19-1	01-20-1	01-27-1	0.5	6.1	6.6	101
0196	2222	40	60	01-20-1	01-22-1	01-29-1	4.0	60.0	64.0	120
0210	1037	50	150	01-21-1	01-22-1	02-02-1	1.5	15.0	16.5	126
0237	8966	20	45	01-22-1	02-02-1	02-02-1	1.0	4.0	5.0	134
0244	7659	70	20	01-22-1	02-02-1	02-04-1	0.5	1.0	1.5	115
0259	4093	30	60	02-02-1	02-04-1	02-10-1	2.0	15.0	17.0	120

WC TOTAL HOURS:

LATE: 6.6
ON-TIME: 87.0
EARLY: 17.0

DUE NEXT 2 DAYS: 34.5

Fig. 6-6. A typical work center dispatch list.

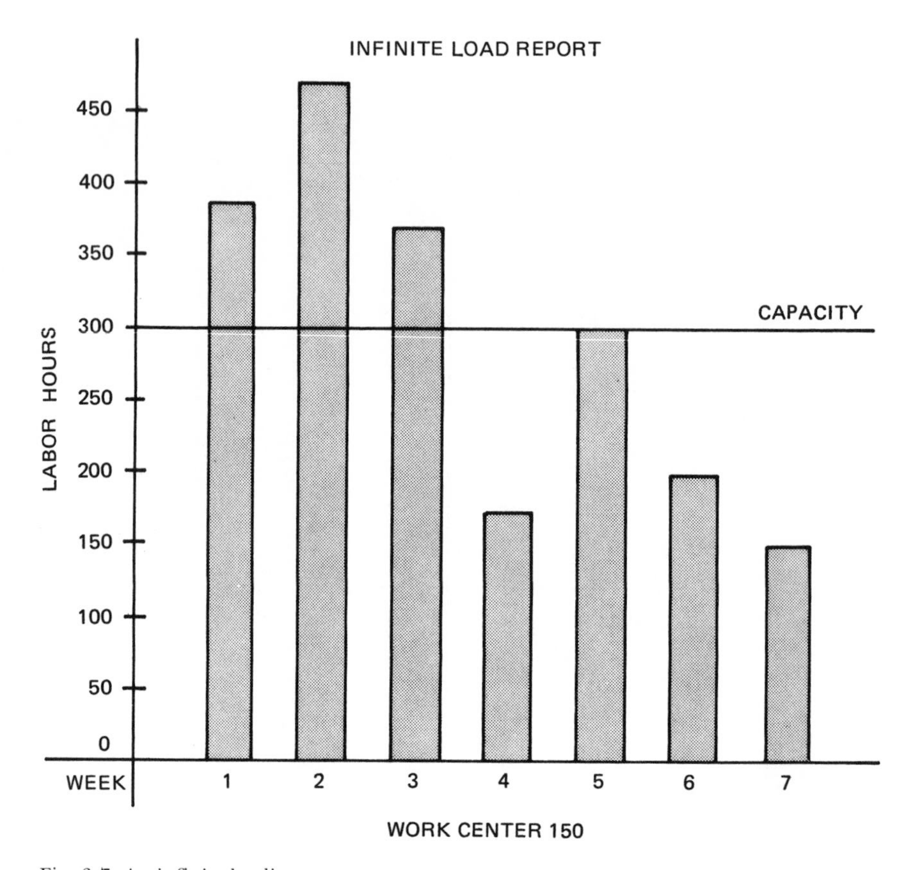

Fig. 6-7. An infinite loading report.

In *finite* loading, each machine or work center has a limit or finite capacity assigned to it. The computer then shifts load around keeping priorities in control so that capacity constraints are not violated (Fig. 6-8).

A major problem with finite loading is that work is continually pushed out on the time horizon if capacity is limited in a work center of the plant. Finite loaders may not show production planners where the capacity problems are so that they can be alleviated.

Finite capacity loading is a complex, time-consuming, and expensive scheduling method that so far has not been of much value in the real world. Countless PhD theses have been written on finite loading with fancy line balancing algorithms and/or linear programs for production scheduling, but so far these have had little practical payoff in today's manufacturing environment. Finite capacity planning is a "stage-2" im-

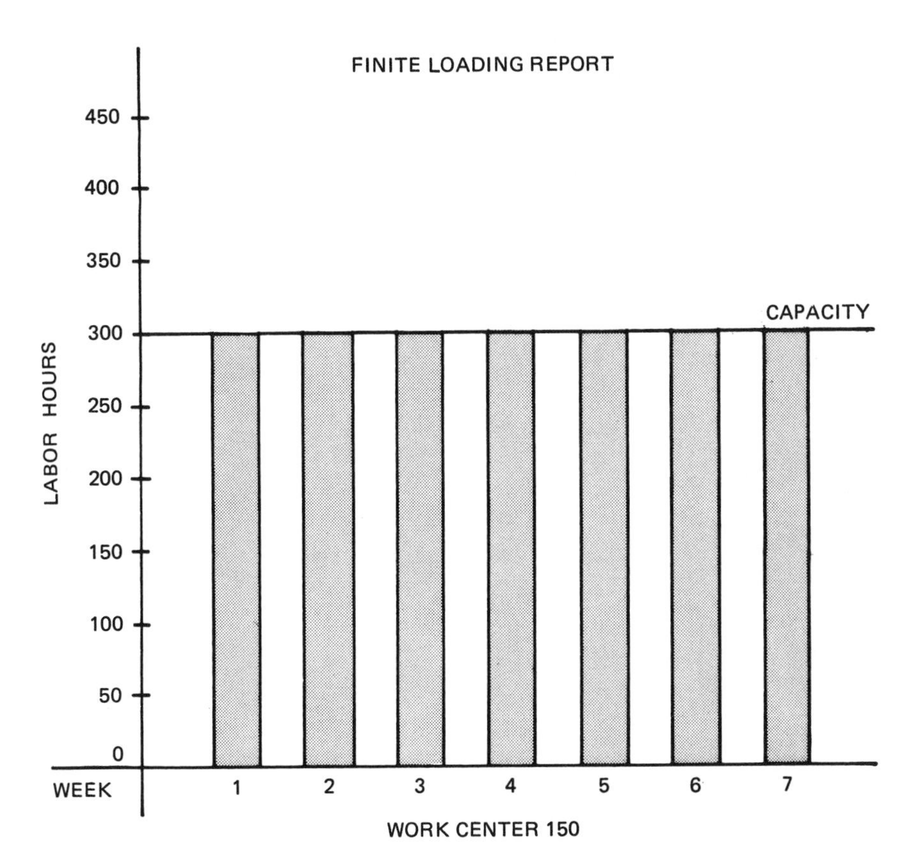

Fig. 6-8. A finite loading report.

provement on MRP. It ought to come into its own in another 10–15 years as computers are able to store more information and operate more quickly and cheaply, and as real-time shop floor feedback control is accomplished through greater use of microprocessors and interactive computer systems. Before this occurs, though, companies can accomplish the primary and most rewarding task of getting a fully operational "class A" closed-loop MRP system installed and running smoothly.

Production management is continuing to gain experience in quantifying what is supposed to happen on the shop floor by day and/or hour, and is getting better and faster at being able to respond to what is happening on the shop floor. Soon, powerful simulation programs will be able to project not only the consequences of present circumstances on the shop floor, but the consequences of *proposed* solutions or "what if" questions as to how to get back on schedule. This simulation of daily operational scheduling and balancing of capacity—like finite loading, I think—is a "stage-2" solution to the real-world problems of today, and won't be available for several years.

The issue of coping with daily capacity problems on the shop floor leads us directly into the next chapter on shop floor control.

7

Shop Floor Control

Shop floor control—what does it mean? Shop floor implies activity out in the production area—where we actually find the work in process. *Control* is the operative word, however, and it is the one word that we will concentrate on in this chapter.

Control Mechanisms

The word control cannot be discussed without considering the three necessary features of any control mechanism. These are:

1 / *Standards.* A standard to control to, i.e., a goal, usually expressed as a number plus and/or minus an allowable deviation from this goal.

2 / *Feedback.* A feedback loop or mechanism that indicates what is happening in the real environment, and how that compares with the established standard.

3 / *Correction.* A correction or reaction mechanism that will attempt to return an out of control situation back to normal.

The concept of control outlined above is pictured in Fig. 7-1.

Priority Control

What are we trying to control on the shop floor? Manufacturing PRIORITIES! Priorities come from the master production schedule and mrp output, which show in order of need items to be manufactured. The master production schedule is the production schedule which has set top-level priorities according to Marketing's needs; mrp creates prioritized planned orders based on master production schedule requirements. Once these orders are released, shop floor control must be sure that the priorities on the shop orders are met. Thus, shop floor control is the feedback step that "closes the loop" to make an MRP system a "class A" system as shown in Fig. 7-2.

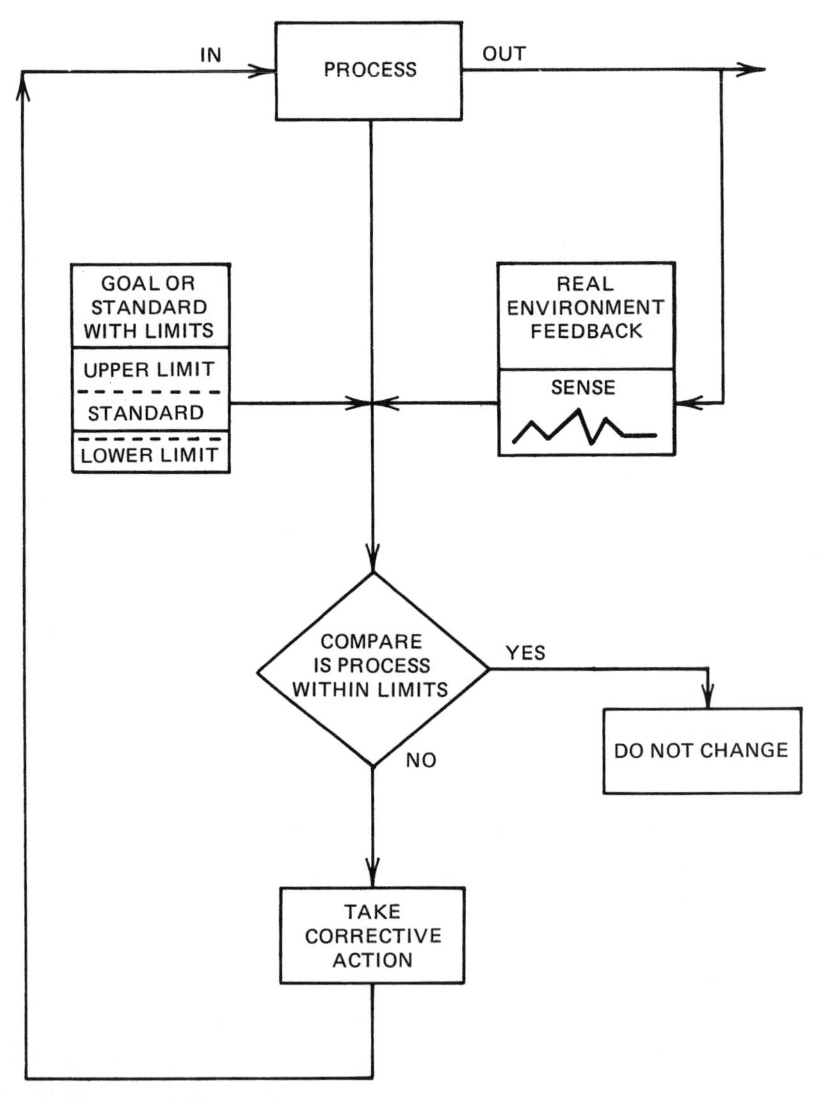

Fig. 7-1. The concept of control.

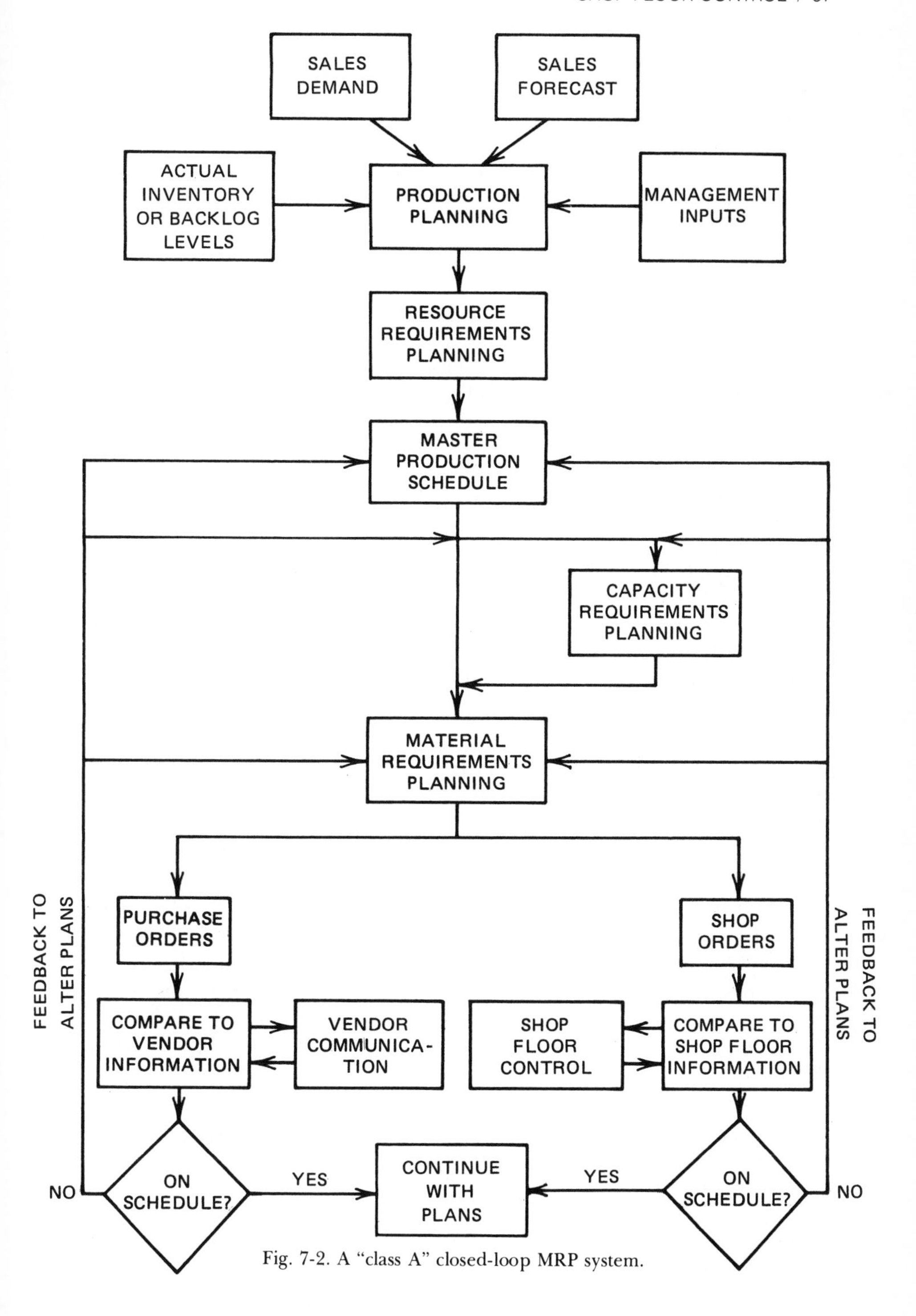

Fig. 7-2. A "class A" closed-loop MRP system.

Information Needed from Shop Floor Control

The questions shop floor control has to answer are:

1 / Are the correct quantity and types of items being manufactured as planned by the master production schedule and material requirements plan?

2 / Is the shop on schedule as dictated by mrp due dates and priorities, or is it gaining, or losing?

3 / If the shop is behind schedule, what has to be done to get it back on schedule?

4 / What is the current status of any shop order or part on the production floor? Where is the order, and is it there at the right time?

These questions and the *control* of work in process are both related to two principal topics—lead time control and input/output control.

The Lead Time Syndrome

As we discussed in Chapter 4, the entire requirements time-phasing concept of master scheduling and mrp is totally dependent on lead times. Thus, to keep the system priorities correct, control of lead times is vital. The key to lead time control is an understanding of the lead time syndrome.

The lead time syndrome arises from the fallacious but common rule that if products are late getting through production, lead times should be increased to allow for this. In this way, according to this theory, marketing people will have a more realistic picture of true manufacturing lead time. Not only is this untrue, but if this rule of thumb is followed, it will only confirm Parkinson's law that work in process expands to fill the space or time allocated to it!

Here is how the lead time syndrome really works. Assume that one of your vendors is quoting a lead time of 6 weeks, but is constantly behind in delivering the product. Let us further say that you order 10 units a week from him to cover your weekly requirements. At the outset then, the situation is this:

Lead time	Units on order	Vendor backlog
6 weeks	60	60

Now, the vendor, knowing he owes you 60 units and is behind, finally "acknowledges reality" and moves his lead time up to an "honest" 10 weeks in order (he thinks) to have a better chance of making on-time deliveries. What is your first reaction? You automatically increase your orders to cover the 10-week period instead of the 6-week period, by sending the vendor 4 more weeks worth of orders. Now the situation looks like this:

	Lead time	Units on order	Vendor backlog
Before	6	60	60
Now	10	100	100

Note that the vendor is even more behind because the backlog has just gone up by the extra 4 weeks worth of units you need, due to the new 10-week lead time. Now, to relieve some more pressure, the vendor goes to a 12-week lead time! The result is that you order 2 more weeks worth of materials! This cycle can go on and on! Note two things, though:

1 / The vendor only compounds the problem—the longer the lead time, the more the backlog automatically rises.

2 / As the lead time increases, the vendor's customers have to estimate their order needs even further out on the time horizon—where their forecast becomes increasingly unreliable. Hence the tendency on the part of the orderer, especially in a tight supply situation, is to overbuy in order to protect against stockouts. The fact that the long-term forecast is less accurate both in quantity and as to which product is needed also guarantees that as time progresses toward those far off weeks, order changes or cancellations will increase drastically as the forecast becomes more near term and accurate, or rather the original forecast proves increasingly unreliable. Eventually, the whole system is over burdened with orders that are no longer needed but were not cancelled, and new orders piled on top of old ones. As a result, vendor capacity is over bought.

Now, assume the vendor increases capacity by adding labor or by some other means. Immediately, the vendor can announce to the salesmen that they can start quoting 8-week lead times instead of 12 weeks. What happens? You, already having 12 week's needs on order, do not need to order anything for 4 weeks! The vendor, seeing orders dry up for the 4 week period, panics and concludes falsely that competition's lead time is lower, and thus lowers his lead time again to remain competitive, thus allowing you the luxury of not having to order again for a week or two! Well, this cycle can go on and on also! The fact is that lead time is often a direct function of what you allow it to be, or say it is.

The Importance of Accurate Lead Times

We have seen how mrp's scheduling algorithm backschedules from the part's due date based on the operation times on the routing and the lot size of the part. In addition, we have just seen how the lead time syndrome can disrupt the scheduling process. These two subject areas continue to be the most misunderstood or underappreciated areas in MRP. Few manufacturing establishments run their operations and MRP systems based on accurate lead times.

Many companies have poor industrial engineering standards on their part routings. While these setup and run standards can vary high or low, they invariably are higher than necessary. In addition, the universal fudge factors of queue time and move times are often grossly distorted. I have seen parts that took only 3 days to make but that had 17 days move and/or queue time in their routings. This lead time inflation has a disasterous effect on an MRP system because it destroys all priorities needed to keep the due dates valid. Inflated lead times "pull in" part orders that are really not needed. As a result work in process is increased to the point where the shop floor is choked with parts, priority control is lost (or never gained), and everyone blames the problem on the computer!

As an example (Fig. 7-3), let us see what inflated lead times do to mrp. Observe that for a part whose due date is week 9, inflated lead times will cause the part to be started 4 weeks earlier than really necessary! Assume now that you have another part with an accurate lead time totaling 8 weeks. In mrp, both parts would have equal priorities if the

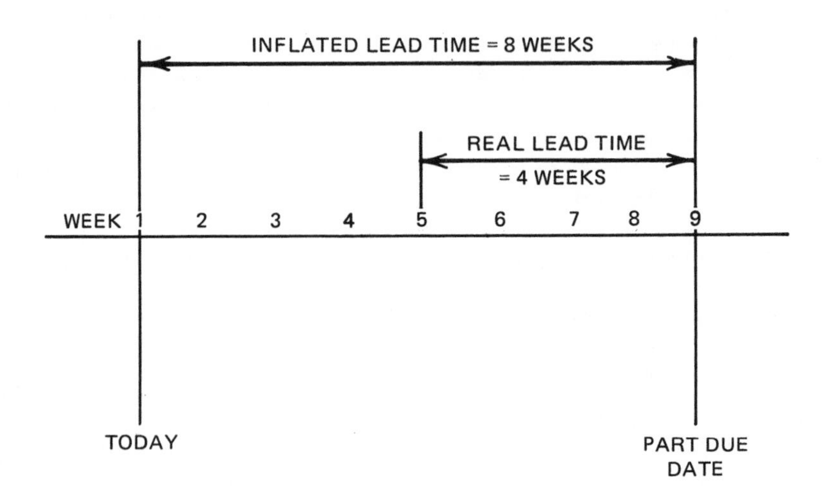

Fig. 7-3. mrp with an inflated lead time.

inflated lead time was utilized on our first part. This is wrong, but only the informal system—not mrp reports—could sort out what the true priorities on these parts should be.

Many companies try to get away with just a little "fat" on their lead times. *Any* fat destroys mrp priorities and will degrade the system. Even small lead time excesses can compound priority problems, especially in bucketed mrp systems. Just hours of extra lead time can cause a part's start date to be rounded off to an earlier week, thus moving the order an entire (unnecessary) week.

Correct lead times are vital to maintaining proper shop floor priorities and to having everyone trust the system. Note, I am not advocationg making standards unfair or unreasonable—they must be accurate, though. If all standards are loose, then they should be made realistic so that the factory can be scheduled correctly. Usually, workers gain time on some jobs and lose on others. Concurrently, the entire wage compensation system may have to be reviewed in order to get the pay scales adjusted properly. Production scheduling is one problem. Worker's wage rates are another distinct problem.

Standards and, indeed, routings as a whole, are very difficult to stay on "top of." New plant floor layouts, new machine tools, and a constant flow of engineering changes mean continual changes to routing operations and times. Often the implementation or use of an MRP system will expose poor industrial engineering standards and an Industrial Engineering department that is understaffed and way behind in its work. This situation is intolerable. The success of a company's MRP installation is directly related to the quality of its manufacturing time standards and Industrial Engineering department. If your company's standards are correct but you are concerned that lead times may be excessive, gradually and periodically tighten them by cutting down on move and queue times. If they become too tight, something will "squeak," and you will note a problem on an individual work center's input/output reports. As long as there are no "squeaks," keep selectively tightening the move and queue times in a common sense manner. You will soon see the difference in decreased work in process and truer shop floor priorities.

Input/Output Controls

The concept of input/output (I/O) control can be used on any measurable subpart of the production process—from one machine or work station to a department or room, to the entire production floor. We treat whatever we are controlling as a "black box" into which we input something to be worked on, and a (more) finished product is outputted, as shown in Fig. 7-4.

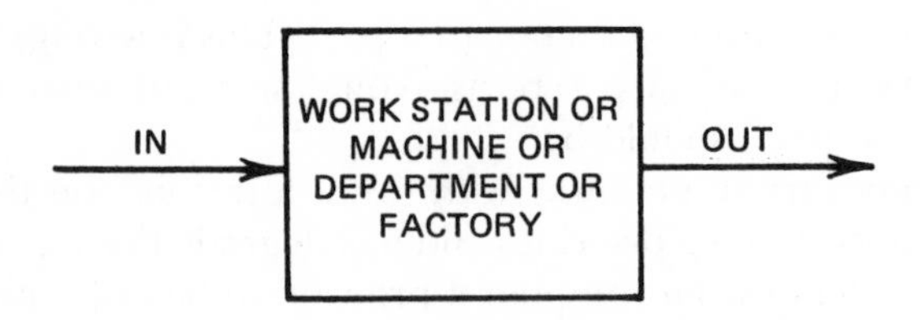

Fig. 7-4. Basic input/output.

Assume that we have examined the input rates (per hour, day, week, etc.) and output rates of an assembly line and have the data shown in Fig. 7-5 for 10 days. Now subtract input from output to calculate the difference on a daily basis. Output minus input means that for a situation where output is less than input, the resulting answer's sign will be negative. This relates to the fact that ground was *lost* against the schedule. At the same time also calculate the cumulative I/O difference starting with the backlog of 16 units. The results are shown in Fig. 7-6. Then graph both the daily I/O differences and the cumulative I/O differences. The daily I/O difference graph is shown in Fig. 7-7. Note that the reference line for the graph is *zero,* that is, where output equals input.

The graph of the cumulative I/O differences, starting with the backlog entering day 1, is shown in Fig. 7-8. Here, in the cumulative graph, the reference line zero can represent either zero backlog or work in process; or in some industries or for some types of products, it can represent a "standard" work in process. Therefore, this cumulative I/O graph can

	Units	
Day	In	Out
1	20	16
2	25	18
3	15	20
4	30	20
5	30	25
6	25	25
7	25	22
8	15	22
9	10	18
10	30	20

Fig. 7-5. Sample input/output rates. (Backlog at start of day 1 is 16 units.)

Day	Units In	Units Out	Out−In	Cumulative Out−In −16 beginning backlog or WIP
1	20	16	− 4	−20
2	25	18	− 7	−27
3	15	20	+ 5	−22
4	30	20	−10	−32
5	30	25	− 5	−37
6	25	25	0	−37
7	25	22	− 3	−40
8	15	22	+ 7	−33
9	10	18	+ 8	−25
10	30	20	−10	−35

Fig. 7-6. Sample input/output calculation.

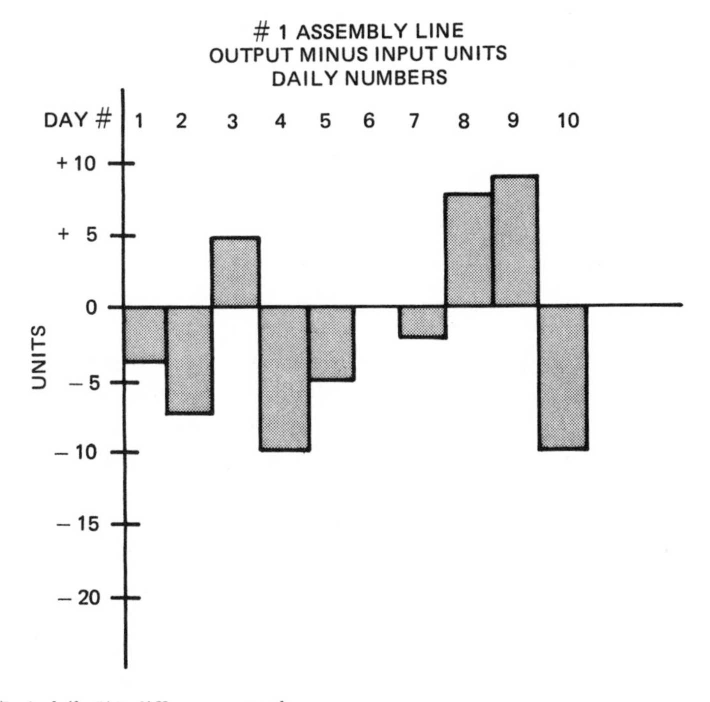

Fig. 7-7. A daily I/O difference graph.

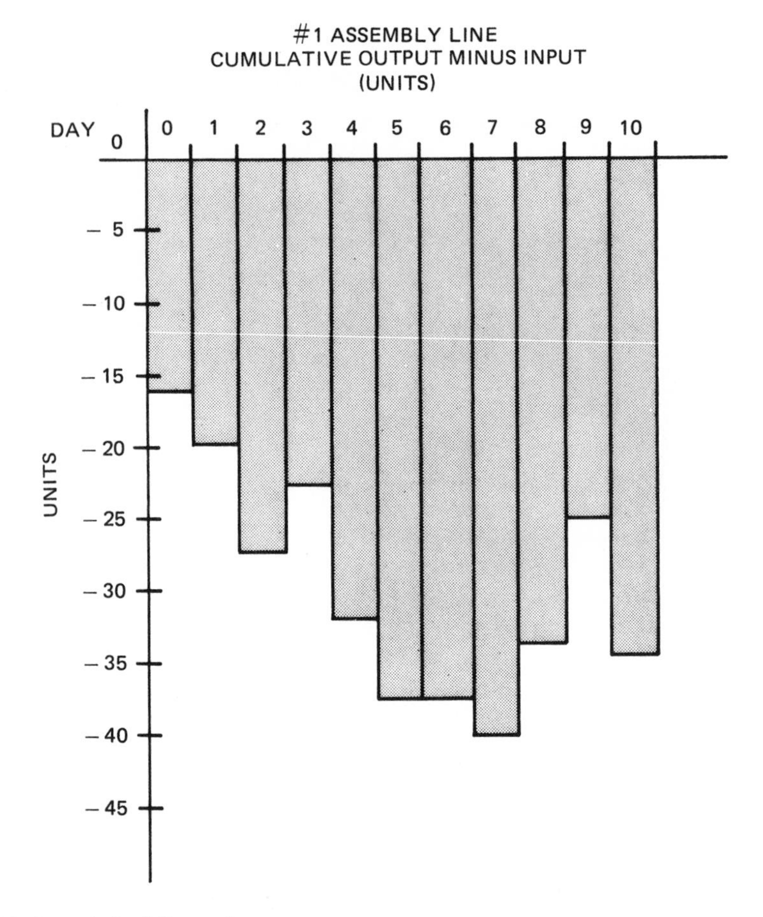

Fig. 7-8. Cumulative I/O graph.

easily show total backlog or work in process as shown previously. The standard work in process *plus backlog* could just as easily have been illustrated if we placed the zero reference at −16 units. In any event, these cumulative graphs show the daily effects of input/output on the total manufacturing work load.

Note that only on days where output minus input was *positive* did we gain on the work in process or backlog situation. On days were output minus input was negative, all we did was to increase our backlog! Tolerances or limits can be set on the daily I/O differences as shown in Fig. 7-9. Once these tolerance limits are established, a clerk or the computer can be assigned to warn management anytime the daily I/O difference exceeds these limits. Then management can take steps to remedy the

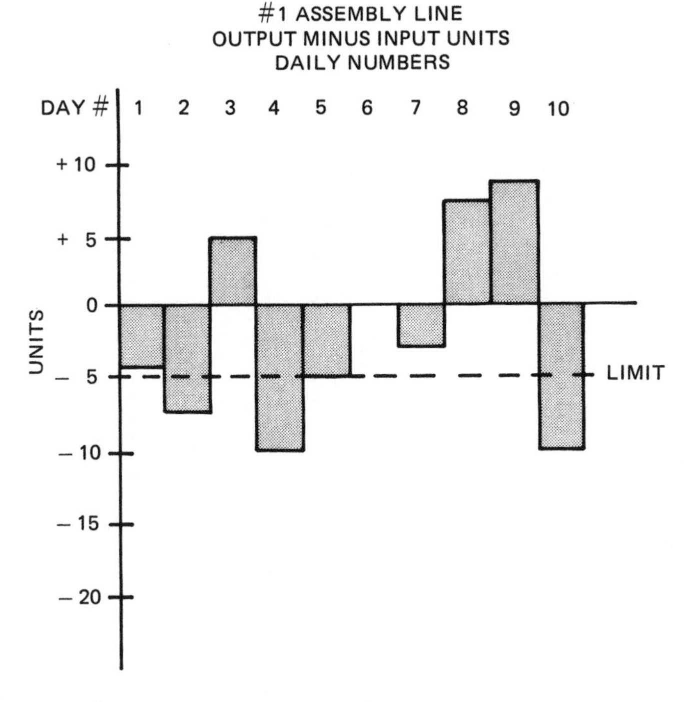

Fig. 7-9. Daily I/O difference graph with limits.

problem immediately, before the backlog or work in process can increase substantially or before it reaches such a low point that production workers run out of work.

The same type of control limit could be shown on the cumulative input/output graph where the base of zero was set as no work in process at all, and the control line was the standard or desired work in process, as shown in Fig. 7-10. The entire point of using the I/O controls is that *input should never exceed output!* If it does, you are only misleading yourself, filling up the production floor with more material, and increasing the confusion on order tracking and priorities.

What often occurs in a work center or plant is that when input equals output, output is not high enough! The *real problem,* then, is one of inadequate *capacity*—and the answer is not more input but first more throughput capacity. Some management people's response to low output is to just keep piling orders into work in process. This is an abdication of their management responsibilities, and unlike the liquid analogy (Fig. 7-11) used to show work in process and capacity restrictions, increasing the liquid level or pressure will *not* increase the output!

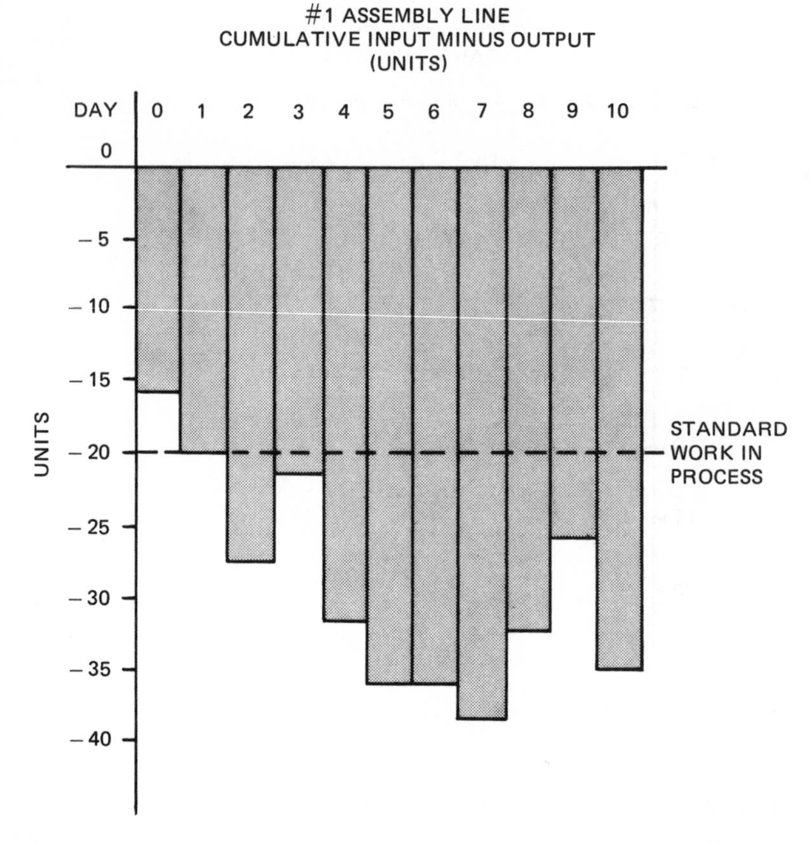

Fig. 7-10. Cumulative I/O graph with control limit.

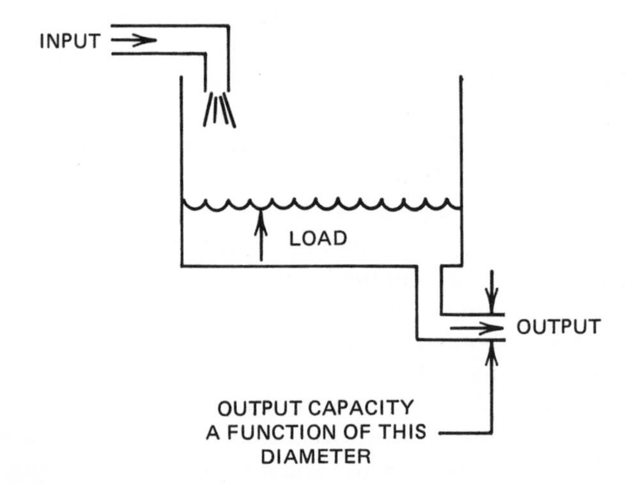

Fig. 7-11. The liquid analogy of load and capacity.

The output necessary should be planned and leveled at the master production schedule level. The key to controlling the shop floor is leveling the work input, and controlling the input to achieve the output desired. Input/output controls are so simple and effective to use that it is a pity so many managers ignore this basic and important control technique.

Load Leveling with Firm Planned Orders

mrp has a unique ability to report out planned hours by work center to the limits of the master production schedule planning horizon. This allows manufacturing planners to spot overcapacity work center loads in sufficient time to correct the problem by leveling input to the work center. Figure 7-12 illustrates a typical problem.

Planned Hours Report for WC 150

100% capacity for WC 150 = 200 hours

Week	1	2	3	4	5	6	7	8
Planned hours	200	204	197	180	170	250	150	180

Fig. 7-12. A planned hours report showing an overload.

The planned order requirement in week 6 is 50 hours over capacity. Note, however, that there is excess capacity available in weeks 5 and 7 on either side of week 6. The technique that will enable us to move some of those week 6 planned hours is the use of a *firm planned order.* In order to use this tool, though, we need to find the parts that are causing the lump of 250 planned hours in week 6. This is done by *pegging.*

Pegging

Pegging is being able to tie material or labor hour requirements to either the *part or order* that caused them. In the *pegging* process, we proceed up the product structure level by level until we have found all the part orders causing the 250 labor hour requirement. This level-by-level process is called single-level pegging.

Full pegging refers to the computer's ability to trace and print out in one report all the level-by-level requirements all the way up to the (0 or 1 level) item that is master scheduled. Full pegging is not really needed, however, since the planner should look for a solution at each bill of materials level. Furthermore, full pegging promotes the tendency by the planner to proceed straight to the top level and change the master production schedule. This action should only be carried as a last resort.

Another, longer way to find the parts generating the 250 hours of labor requirements is to use the where-used report on a level-by-level search. This process is extremely time consuming when a part, for instance, can be used in 30 parent parts. Which of the 30 parent parts has an order in WC 150 in week 6? Then find all the rest of them!

Single-level pegging would produce a report that would detail week 6's labor requirements in WC 150 (Fig. 7-13).

Week 6 - WC 150

Part number	Planned labor (hours)
1946	48
7328	83
3921	75
4425	44
	250

Fig. 7-13. A single-level pegging report.

The planner could then easily select part #1946 or #4425 and move that part into week 7 using a firm planned order.

Firm Planned Orders

Normally, mrp logic will juggle planned order due dates and order quantities around to accommodate changes in the master production schedule, on-hand inventory balances, part allocations, safety stock requirements, and scheduled receipts of open orders.

Firm planned orders give the planner a method to *fix* the order quantity or due date or start date so that the mrp logic cannot automatically change any of these items. In other words, by fixing the order due date and quantity, mrp will have to work other requirements around the firm planned order. The prime use of firm planned orders is in just the situation outlined earlier. There, we could create a firm planned order for Part #1946, whose due date would be in week 7 instead of week 6, and whose quantity would be fixed at 48. Thus, week 6's labor hour requirements are reduced to 202, and the weekly sequence of planned hours for WC 150 is now shown in Fig. 7-14.

Planned Hours Report for WC 150

100% capacity for WC 150 = 200 hours

Week	1	2	3	4	5	6	7	8
Planned hours	200	204	197	180	170	202	198	180

Fig. 7-14. A revised planned hours report after using the firm planned order.

The major caveat concerning firm planned orders is that the technique should be used sparingly. If not, all orders will become firm planned and the mrp logic will be rendered useless.

Shop Floor Status

Order or lot tracking through the production system traditionally has been one of the toughest parts of manufacturing. Indeed, in many companies, a request for order or lot status inevitably causes a frantic search of the shop floor for the required order. Especially in today's computer environment, this procedure is no longer necessary. An order or lot status request asks two basic questions:

1 / Where—that is, in what department or work center—is the order?

2 / Is the order on schedule?

The first question can be answered in a number of ways.

The oldest and simplest method to track order status is to have a set of cards or tear off tags (now computer printed) that accompany the lot or order through the production area. When a lot or order clears a department, or work center, the tag or card is handed in to Production Control, where the order's status is posted as being in the next department on the order's routing. These status cards can also contain "move" instruction cards that direct material handlers as to where to move the order after each operation is accomplished.

A more modern technique for order tracking is to have these computer printed cards or coupons as part of the routing. When they are turned into Production Control, they are then scanned by an optical scanner for daily updates to computerized work in process tracking system.

The problem with these two ideas is one of how to minimize the time lags between reporting and lot/order movement, especially in plants where the entire manufacturing cycle time is short. The answer for this problem is to have real-time manufacturing-data-base update where, through keyboards similar to a calculator or through cathode ray tubes (CRTs), either located right on the shop floor, workers enter case numbers, lot numbers, order numbers, serial numbers, etc., as the work enters and leaves their area. This assures real-time work in process tracking, and leaves an audit trail to track missing work in process. Never has this real-time work in process system been more feasible than today, what with the tremendous growth of portable small data entry terminals, Universal Product Code (UPC) or bar code readers, and inexpensive minicomputers.

Asking if the lot or order is on schedule implies that you have a schedule firmly established against which to track manufacturing progress. If so, it is easy to load this schedule into a computer, and use real-time floor updating to generate exception reports indicating late orders. These late orders can then be given the attention they need to get them back on schedule.

The Human Factor in Shop Floor Control

Many firms ignore the human factor in putting in shop floor controls and in attempting to reduce backlogs and lead times. Remember that a large backlog on the production floor represents the production worker's *security*—the promise of work for the future. This is particulary so if the worker is on piece rate or on incentive bonus. These workers will naturally resist all management efforts to reduce the work in process and backlog. Management, therefore, must take appropriate steps to guarantee that there will be sufficient work for them in the future, and that the workers will not suffer any loss in pay if—due to a scheduling problem—the work "dries up" on the shop floor. It should be far cheaper for management to make this guarantee than to run manufacturing with the inflated work in process and inventories that some of them do today.

Management must also make a constant effort to educate the production workers concerning the new manufacturing systems being put in place, the worker's vital role in making that system work properly, and the fact that if there are any temporary disruptions in the system, the workers will not pay a penalty for it. More discussion of these points will be given in Chapter 11.

Another caution is to make sure that company personnel have adequate information with regard to what is happening on the manufacturing floor. In many companies when things are not going smoothly, management reports tend to be omitted or "sat on" until the situation improves. Many times the lack of information concerning the fact that something is wrong somewhere prevents appropriate and timely curative action from being initiated. No matter what the situation on the manufacturing floor, production control reports must be accurate, complete, and issued regularly.

The Foreman's Role in Shop Floor Control

One of the main benefits of an MRP system is that it allows the foreman to be a foreman! Many studies have shown that in the informal pre-MRP world, a foreman spends a majority of his or her time chasing part shortages, expediting the "hot job" of the hour, and attending meetings to find out what he or she is supposed to be working on.

In the MRP world, valid work priorities are established and maintained by the system and good shop floor feedback. This frees the foreman of previous time-consuming and frantic attempts to stay on top of things, and allows him or her to concentrate on being the leader of a group of people engaged in a common task. Three of the "new" duties the foremen can pursue are:

> The education and training of people about their role in their MRP environment. Not only must the people be trained—or told and shown how to do the job—but they should be educated as to why their job is important and why it is important for the job to be done right the first time.
>
> Concentrating on improving the manufacturing process in his or her own work center. This could involve such tasks as improving tools, jigs, and fixtures; creating new part racks; cleaning up and painting the area; and asking his or her people for their suggestions on needed improvements.
>
> Upgrading his or her own skills—either in technical or leadership or company-oriented management areas.

These are the real jobs of foremen and supervisors. Instead of being glorified expediters, with MRP they can at last become true shop floor managers for the company.

8

Purchasing

One of the best benefits of manufacturing resource planning systems is that they finally allow the purchasing people to do their job—purchasing, as opposed to "fire fighting" and worrying about inventory control problems. Because of its fundamental importance to the integrity of the MRP system, raw material and parts inventory control today is usually a separate function that often reports directly to the Vice President of Manufacturing or Material Manager.

In the manufacturing planning and control system environment, the Purchasing department usually works with three or four groups of data of its own, in addition to the material master data and the raw material/parts inventory data. Three necessary data groups are

1 / Open purchase order data

2 / Purchase order history data

3 / Vendor specification data

These three data groups form the basis for a number of reports used by the purchasing agent or buyer.

Open Purchase Order Report

Open purchase order data contain information concerning all open purchase orders, in effect at any given time.

The purchasing agent or buyer can call out reports of all open purchase orders by scheduled receipt (due) date, order date, part number, vendor name, purchase order number, or total order dollar value. Reports showing all purchase orders open over 30, 60, or 90 days can be obtained for review (Fig. 8-1).

		Dollar Amount				
Purchase order #	Vendor #	30 days	30-60 days	60-90 days	Over 90 days	Total
0157	22	1770	385			2155
0374	10			9375		9375
0732	8				410	410
1533	12	3615	795			4410
3441	5				4215	4215
4009	11	5560	3210	1615	460	10845
		10945	4390	10990	5085	31410

Fig. 8-1. An aged open purchase order report.

Exception reporting of past-due scheduled receipts (late orders) based on this open purchase order data is possible. In addition, all purchase orders scheduled for receipt within 1 week, for instance, can be listed as a reminder to purchasing planners to check on whether vendors will meet their promised shipment dates. Such a report is shown in Fig. 8-2.

For the purchasing agent or buyer, the information based on these open purchase order data is their day-to-day "bible". As with any data base, it must be kept accurate and current to be effective.

Purchase Order History Reports

Purchase order history data contain essentially the same information as the open purchase order file but also include data on the order receipt date, part of material price, freight charge, and perhaps even some indicator as to what extent the incoming parts passed quality control inspection. These data can be sorted and called out in almost any manner, as illustrated in Fig. 8-3. They form an invaluable basis for real purchasing research, vendor rating, and, of course, a purchasing audit trail.

The following purchase orders are due in next week, i.e., from 09-08-80 to 09-12-80:

Purchase order #	Vendor #	Vendor name	Vendor phone number	Part #	Part description	Quantity	Dollar amount
0169	15	Atlas	617-549-1237	76-0015	Bolt	100	115.05
0375	9	Universal	213-659-4522	49-1052	Tubing	70	475.00
1572	4	U.S.M.	401-732-1311	70-9906	Valve	15	1732.00
2001	7	Jones	202-666-1573	32-8846	Spring	10	56.00

Fig. 8-2. Purchase-orders-due-in-next-week report.

Purchase Order History

Part # 76-5004

Capscrew, Allen Head

⅜" × 24 × 2"

Black Anodized

Purchase order		Vendor			Unit of	Price	Q/C	Lead time
#	Date	#	Name	Quantity	measure	per U/M	accept	weeks
504	6/15/78	17	Atlas	100	each	0.75	Y	2
691	10/2/78	22	Jones	150	each	0.76	Y	3
743	1/17/79	17	Atlas	100	each	0.79	Y	2
967	4/13/79	22	Jones	100	each	0.82	N	3
1004	4/29/79	72	Smith	150	each	0.83	Y	2
1276	6/12/79	17	Atlas	100	each	0.83	Y	2

Fig. 8-3. Purchase order history for a part.

Vendor Reports

Vendor data would include a complete listing of all current and past (back to some arbitrary point) vendors used by the company. These would include data such as vendor number, vendor name and address, billing terms, minimum order amount, key salesmen or management contact information, and possible current and past year or month to date purchase dollar volume. Usually reports based on these data are listed out by vendor number. Information from vendor reports could be used to show an ABC analysis of year-to-date purchases from vendor accounts, for instance. A typical vendor report is shown in Fig. 8-4.

Vendor Rating Systems

A fourth data group that can be useful is one on which a formal vendor rating package can be based. This package may be in-house designed, or it might be purchased from a software supplier. Using a vendor rating system that is based on careful consideration of factors important to your firm and industry, vendors are rated on their ability to deliver on at least the three most important factors of

Price
Quality
Timeliness or service

Other rating factors that might be considered are shipping costs, packaging quality, product appearance, product safety/liability considerations, etc. (Fig. 8-5). The vendor rating package can be an important guide for choosing between alternative suppliers. Just as important, it can also highlight situations where the vendor you are now using is not competitive, and you should be looking for a second source!

Vendor Communication Using mrp

The key feature of mrp for buyers or purchasing agents is that it allows them to establish *reliable communication* with their vendors—laying out future purchasing requirements in a time-phased manner. Vendors, instead of always being in a reactionary mode trying "to get out every order yesterday," can then, perhaps for the first time, do some future capacity and financial planning of their own.

Vendor Listing Report

Date: May 31, 1979

#	Name	Address	Phone number	Contact	Minimum order	$ Volume last year	$ Volume year to date
0179	Jones Supply	16 Deer St Avondale, NY 11732	617-542-7743	Jim Shultz	$50	17040	5736
0242	Atlas Distributors	317 Park St Itaka, MD 40342	210-777-1009	Mary Smith	$100	4732	1004
0731	D G Supply	4445 Broadway Chicago, IL 60735	312-913-0415	George Rowan	$150	50755	32010

Fig. 8-4. A vendor listing report.

Vendor Rating Report
Vendor: #0734
Atlas Supply Corp
1779 Jericho Road
Huntington, WV
37205

% Orders low price	Actual lead time to quote ratio	% Orders passing Q/C	Ship cost per order	Order fill ratio % of line items	Average price change YTD
36%	0.96	99%	7.36	97%	+8%

Fig. 8-5. A vendor rating report.

De-expediting with mrp

Another unique feature of mrp is that it tells buyers or the purchasing agents when they can *de-expedite* an order—that is, allow the vendor *more* time to get the order shipped. Vendors seldom get such requests; they usually only receive expedite calls or cancellations. Often, this de-expedite knowledge can be helpful when the buyer needs another order expedited. Just as in a fixed-capacity situation if one order is to be expedited, another must be de-expedited. With mrp, a buyer has the information to make the right priority choice between parts and to communicate this to the vendor.

Controlling Vendor Lead Times

As we have seen in Chapter 7, the control of lead times is crucial to the proper operation of any mrp system. It is up to Industrial Engineering to see that Manufacturing has accurate time standards on which to base mrp scheduling. By the same reasoning, the responsibility for accurate lead times on purchased parts rests with the purchasing manager, who must monitor vigorously lead time changes. These purchasing lead times, as with manufacturing lead times, should be an average of past history. I have seen one purchasing operation use generous "Worst case" lead times that destroyed mrp priorities, and allowed part stocks to be excessively high.

Vendor lead times constantly change as industry, national, and international supply and demand conditions change. There can be a switch of vendors as a source for a given part when lead times change or for a variety of other reasons such as changing prices and sourcing specifications. It is convenient for the purchasing manager to have a weekly report of vendor lead time changes (Fig. 8-6) that summarizes their number and the number that increased versus the number that decreased. Again, the human tendency on the part of buyers is to protect themselves and increase lead times, but never to decrease them. The purchasing manager can look for a statistical balance where lead time increases should equal decreases over a time period of a few weeks.

Vendor Lead Time Change Report

for Week of: 09-08-80 to 09-12-80

# of Vendor Lead Time Changes 67

Increases

# of Lead Time Increases: 41 (61.2%)

Average Lead Time Increase 1.7 Weeks

Decreases:

# of Lead Time Decreases 26 (38.8%)

Average Lead Time Decrease 0.7 Weeks

Fig. 8-6. A vendor lead time change report.

All purchase orders should be analyzed to calculate the actual vendor lead time taken on all orders. A report such as that illustrated in Fig. 8-7 can then be prepared that compares a statistical average of actual times to current base vendor lead time and that recommends a lead time change if the difference is statistically significant. The basis for lead time changes then can become historical fact, tempered by a buyer's judgment of the probable future. Lead time monitoring is a constant effort necessitated by the requirements for accurate purchasing priorities.

Purchase	Vendor	Lead time days		Actual minus allowed		Date last
order #	#	allowed	actual	days	%	changed
3372	17	15	17	2	13.3	08-12-80
3586	11	33	32	−1	−3.0	02-22-80
4711	06	40	46	6	15.0	10-17-79
5331	18	10	8	−2	−20.0	05-05-80
6334	23	70	95	25	35.7	07-13-78

Fig. 8-7. A vendor lead time analysis report.

Vendor Qualification

Instead of being constantly in a "fire fight" mode, the purchasing agent or buyer using mrp can get down to his or her real job of picking the best vendors who supply the correct products (note these may not be the best quality available), at the right price (not necessarily the lowest price), and on a timely basis. In addition, he or she is free to investigate new sources for critical sole-sourced items. He or she can also devote time to negotiating with important vendors on long-term buying contracts, blanket orders, and systems contracting. All of the above steps tie into a cost-reduction program that should be a standard procedure in any Purchasing department.

General items a buyer or purchasing agent might examine his or her vendors for are:

the vendor's quality of management;
how the vendor manages his or her inventory;
the vendor's financial stability.

The purchasing agent or buyer is then interested in whether the vendor has the *capacity* to support his or her needs. To establish this, he or she might study the vendor's

financial reports
organizational structure
location—of headquarters and warehouses
facilities—space, material-handling, packaging, and shipping capabilities

inventory—product lines and brands carried
procurement practices
inventory control practices
customer service—catalogs, salesmen's knowledge
markets served—including competitors of the buyer's company
degree of computerization and data-processing capability
ability and desire to establish a *long-term* working relationship with the buyer, the buyer's company, and the buyer's MRP system.

A great many of these areas are subjective, but should be considered in choosing suppliers. It may be impractical to apply these criteria to all suppliers; however, after doing an ABC analysis of the company's suppliers by annual dollar volume, a buyer or purchasing agent might examine the "A" vendors in the above manner.

Purchasing Financial Planning

With a standard cost system in place (that is, with standard costs on bills of materials) and tied to inventory or material master data, the purchasing agent can obtain important forward and backward views of his or her function, and can do real financial planning.

Using material requirements planning, materials requirements can be generated for at least the next quarter, and generally for the next year. These requirements can also, for the purchasing agent's purposes, be summarized in dollars, and thus serve as an approximate *pro forma* purchasing budget for the coming time period. In addition, this budget presents top management with a tool to measure the purchasing agent's performance in a quantitative manner. For the first time, top management has something to which they can hold the purchasing agent accountable! By examining historical price data the purchasing agent can begin to have a reliable estimate of raw material price level changes. When compared to future requirements, these provide a basis for altering top management to key material shortages or price increases that may dictate product design changes, material substitutions, foreign sourcing, or a decision to make—not buy—a part.

How Much Material to Include in the mrp System

The question always arises: How many items, or what level of detail, do we include on bills of materials? Are indirect items like adhesives, paint, staples, tapes, etc., included? My answer is a resounding yes, given that most modern computer systems have the capability to handle the additional storage and processing requirements needed. In my view, there is little excuse for not putting *all* items used in the manufacture of a product on the bill of materials. In fact, in today's energy-short world, one can even calculate the required energy per work station to make a product, put that on the product's bill of materials, and thus be able to ascertain your company's future energy requirements and costs. This "down to the last item" philosophy obviously is not necessary to obtain the main benefits of mrp. However, the more items "on the system," the easier the control process becomes, and the less one has to assemble data from reports that are not part of the overall manufacturing planning, control, and reporting system.

Using the combined information obtained from purchasing history records, *specific* future material requirements, open purchase order reports, and a quantitative vendor rating system, the purchasing agent at last has the tools with which to perform his or her real job: that of being *effective* at *purchasing planning* and *execution,* and purchasing cost reduction.

9

Forecasting

At some point, Manufacturing has to be told how many units of what product to produce. In the past, this information often came from one or two people in the company—usually the President, or Vice President of Marketing or Sales. Their judgment was based on a good knowledge of their business, their competition, their industry, the country's economic outlook, their customers buying habits, what they *thought* their manufacturing divisions could produce, and their own "gut feel" or intuition.

These forecasts usually were subjective and involved very little quantitative analysis of past sales history and industry or economic trends. Currently, we can take advantage of the tremendous data-crunching ability of computers to track individual product sales history and other economic data. In addition, the computer can easily aggregate individual product sales into product line groups or families. Using the computer, we can track current sales against marketing targets, past years's sales, or against a computer-generated statistical sales forecast. All of these tools provide a much faster and more-detailed statistical picture to management people than was previously available. Management, however, still has the responsibility of deciding how many of what product to make, and when to make them.

As was shown in Chapter 5, the sales forecast is a major input to the production plan and, hence, the master schedule; it plays a large role in determining what Manufacturing is to produce and at what time.

The discussion on forecasting will be brief, not because the topic is unimportant, but because forecasting is a complex statistical subject that will be interesting to only a very small subset of this book's audience. In addition, one chapter cannot do justice to the subject, especially if kept in the perspective of this entire book's contents. Many books, of course, have been written on forecasting. For those interested, I recommend S. Makridakis and S.C. Wheelwright, *Forecasting—Methods and Applications,* Wiley, New York, 1978. Its 700 pages superbly cover all aspects of forecasting.

In the context of this book, what is important about forecasting? A brief look at three forecasting methods and some general forecasting tips would be useful for the reader to consider.

Forecasting Methods

I will mention three forecasting methods that might be useful to a company wanting to start using forecasting in their planning. Of these three methods, two of them—exponential smoothing and the Box–Jenkins method—function by analyzing a series of past data observed over time.

All time series can be broken down to four general components—trend, seasonality and cyclicality, and a random error term that is the difference between the sum of the final three terms and actual data.

> *Trend* is a consistent long-term movement in one direction, such as constant 1% price increase per year.
>
> *Seasonality* refers to short-term patterns within the year, such as the majority of a firm's sales occurring every spring.
>
> *Cyclicality* is analogous to seasonality but refers to a longer-term period of fluctuations, such as the well-known business cycle with a recession approximately every 4 years.

These three components are illustrated in Fig. 9-1.

Exponential Smoothing

Exponential smoothing is an extension of the concept of moving averages. In moving averages, time series data are smoothed into a forecast by averaging the data over the last N periods, with equal weight given to each time period's value. A 3 month moving average, for instance, sums the past 3 months data and divides by 3. The next month, the oldest month's data observation is dropped and the latest is added to arrive at a new sum and new average.

Exponential smoothing, on the other hand, credits the latest data observations as having more value, and thus is designed to weight the latest observations more in the averaging technique. The weight given to these latest observations is governed by the size of the alpha (α) factor or smoothing constant. If it is low, little weight is given to the latest observa-

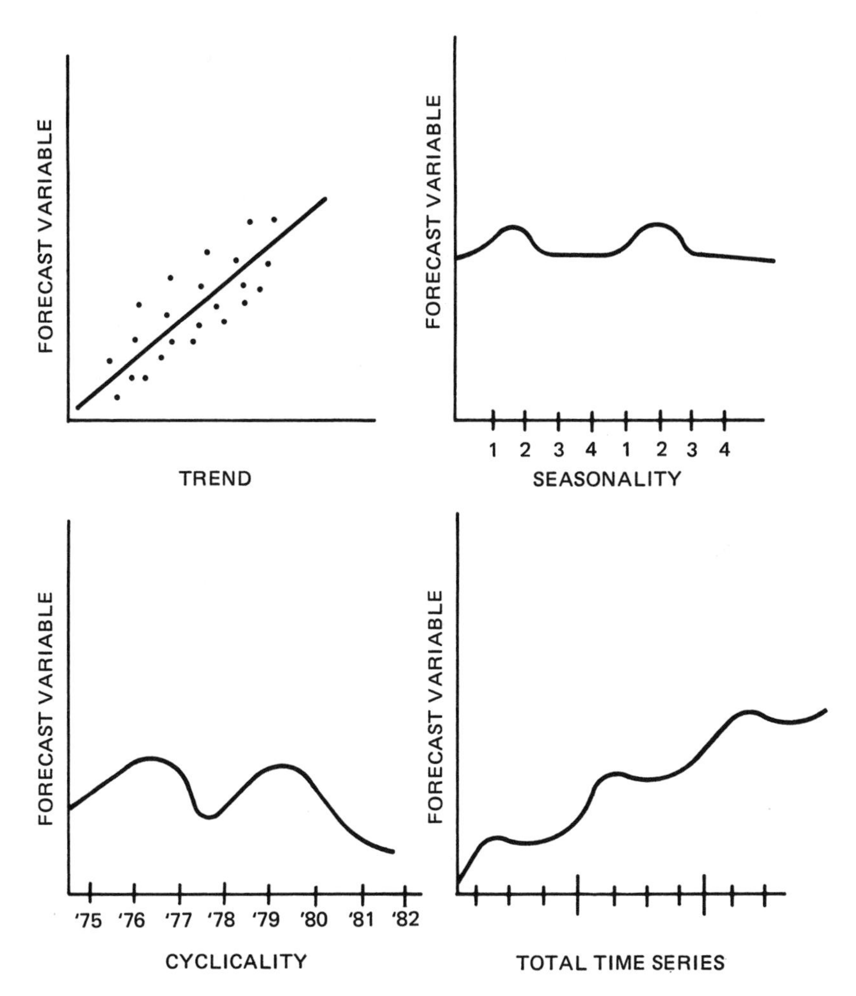

Fig. 9-1. A time series and its components.

tion, and a smoother curve is obtained. Conversely, if the alpha factor is high, a great deal of weight is given to the latest data observation, and the curve obtained is much more erratic. An example of exponential smoothing is shown in Fig. 9-2

Exponential smoothing is easily the most popular forecasting method in use today owing to the fact that it is relatively easy to understand and does not demand much historical data or long computation time on a computer. Many embellishments of exponential smoothing exist, all designed to do a more accurate job of forecasting through better accommodation of trend, seasonality, and cyclicality in the time series to be forecast.

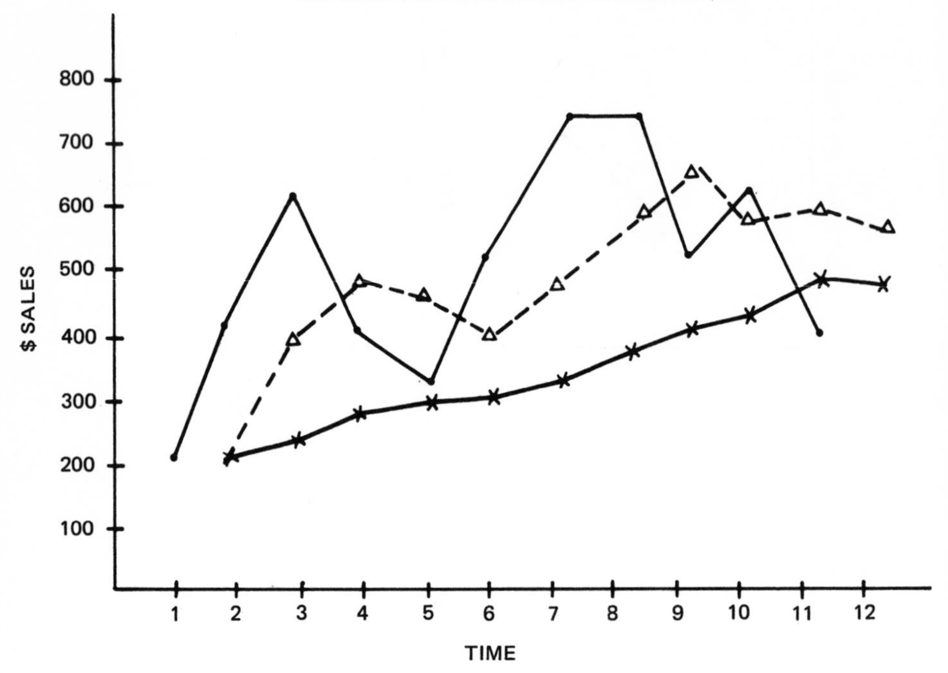

—— • = data

▬▬ × = $\alpha = 0.1$

- - - Δ = $\alpha = 0.4$

$$F_{t+1} = \alpha x_t + (1 - \alpha) F_t$$

where

x_t = actual time period t
F_t = forecast time period t
F_{t+1} = forecast time period $t + 1$
α = smoothing constant

For example, when $\alpha = 0.1$

$$\begin{aligned} F_3 &= \alpha x_2 + (1 - \alpha) F_2 \\ &= 0.1(400) + 0.9(200) \\ &= 40 + 180 \\ &= 220 \end{aligned}$$

Fig. 9-2. An example of exponential smoothing forecasting.

Multiple Regression Analysis

Multiple regression analysis attempts to find *causality* between the dependent variable (sales, for instance) and independent variables such as advertising, price, gross national product level, unemployment, etc.

Multiple regression is based on the easy to understand formula for slope where there is only one variable. This is then called simple linear regression and is illustrated in Fig. 9-3

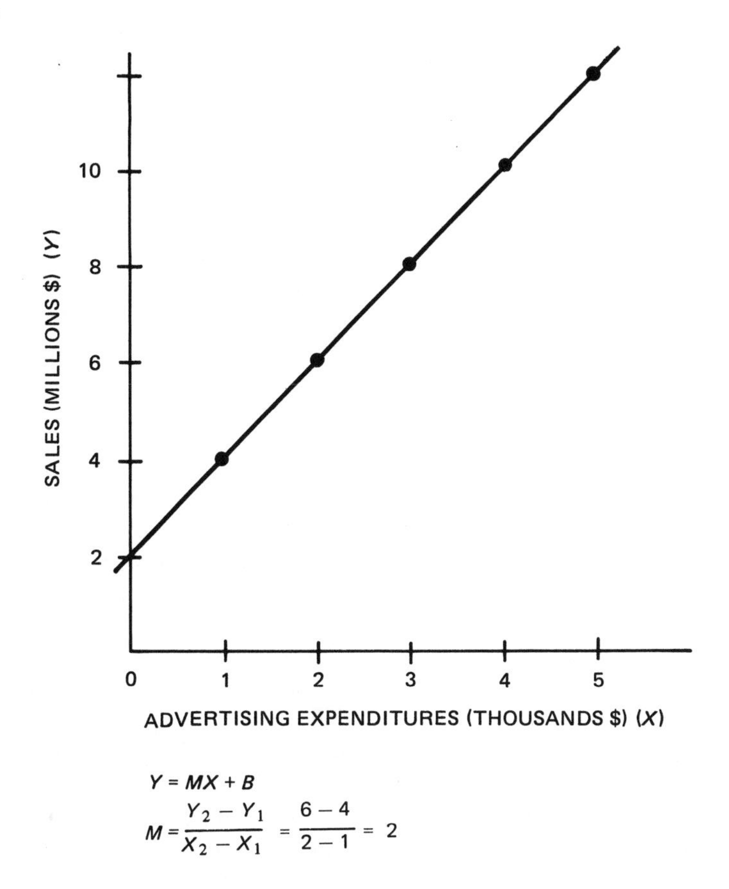

$B = 2$, Y intercept (a constant), therefore,

$$Y = 2X + 2$$

To predict the sales (Y) if \$6000 is spent on advertising,

$$Y = 2(6) + 2$$

$$Y = \$14 \text{ million}$$

Fig. 9-3. An example of simple linear regression forecasting.

In multiple regression, there are more than just the one variable—sometimes as many as 6–10 variables. An example of multiple regression forecasting is shown in Fig. 9-4.

$$Y = a + b_1 X_1 + b_2 X_2 + b_3 X_3 + b_4 X_4$$

where

Y = Sales
a = Constant = 275
b_1 = 0.271
X_1 = Advertising dollars (company)
b_2 = 0.720
X_2 = Personal disposable income
b_3 = 0.094
X_3 = Advertising dollars (industry)
b_4 = - 7180
X_4 = Unemployment

For time period 2,

X_1 = 375,000
X_2 = 10,000
X_3 = 3,540,000
X_4 = 0.08

$$Y = 275 + 0.231(375{,}000) + 0.720(10{,}000) + 0.094(3{,}540{,}000) - 7180(0.08)$$

Y = \$426,286 for time period 2

Fig. 9-4. An example of multiple regression forecasting.

Many companies and industries, after careful study, have been able to isolate variables that influence the sales of their company's products. Once this is accomplished, their multiple regression forecast equation enables them to forecast their sales with surprising accuracy. Regression-based forecasting is not as popular as exponential smoothing. The technique demands that far more data be saved, and further requires careful analytical study of causality that many firms do not have the time or skills to perform. In addition, the computational requirements for multiple regression are greater than for exponential smoothing. While the concept of plugging numbers into a regression forecast is easy to grasp, management often has trouble understanding the statistical interpretation of the regression results. Concepts and terms such as confidence limits, Durbin–Watson statistics, multicollinearity of the data, T-tests, and F-tests overwhelm the average manager who only wants a number.

Box–Jenkins Time Series Analysis

The Box–Jenkins forecasting method has only been in use for 10–15 years. It is an extremely complex forecasting technique based on time series analysis that takes a long time to learn to use effectively. Box–Jenkins attempts to find underlying patterns in time series data that can be used to predict future data points. It can be "tuned" to account for seasonality, trends, and cyclicality. The Box–Jenkins technique requires extremely complicated mathematical procedures first to analyze the existing data, then to develop forecast parameters to account for seasonality, etc., and finally to do the actual forecasting. In addition to this complicated mathematical process, sophisticated human judgment must be interjected in the critical analysis stage of the forecast equation modeling.

While its data requirements are not large, the Box–Jenkins technique requires long computational time to generate a forecast. The vast majority of business managers are not interested enough to ever begin to grasp the theory behind the Box–Jenkins technique. Nonetheless, the Box–Jenkins technique is, if used correctly, a very powerful and accurate method of forecasting.

In choosing a forecasting technique, it is generally easier to select one that management is comfortable with. Most people are not comfortable with numbers for which they have no intuitive feel or working knowledge. This is one of the key reasons why exponential smoothing is so popular and, conversely, why the Box–Jenkins technique will never be popular. Any forecasting technique that a business manager needs a PhD in statistics to operate, interpret, and explain does not stand a chance in the executive suite!

Forecasting Accuracy

There have been many studies performed comparing the forecast accuracy of these three methods. In general, no one technique for forecasting is markedly superior in all cases or even a majority of cases. Statisticians even continue to argue about the best standard of measure to use in determining forecast accuracy. Some common measures of forecast accuracy are the mean squared error (MSE), the common percentage error calculation, and Theil's U-statistic.

Forecasting Costs

Using exponential smoothing as a base, Fig. 9-5 shows a cost index for each of the above methods.

	Program development	Run time
Exponential smoothing	1	1
Multiple regression analysis	6	40
Box–Jenkins	8	176

Fig. 9-5. Cost of forecasting methods—relative dollars required.

Forecasting-Data-Storage Requirements

Again, using exponential smoothing as a base, data-storage requirements are shown in Fig. 9-6 for the three forecasting methods under consideration.

	Program	Data
Exponential smoothing	1	1
Multiple regresson analysis	2.8	10
Box–Jenkins	7.5	24

Fig. 9-6. Data-storage requirements of forecasting methods.

General Observations on Forecasting

The further out into the future you are trying to forecast, the greater the error will be.

Be careful to define the level of data detailed before you start forecasting. Anticipate future products and/or product groupings.

It does not hurt to collect very detailed data. The computer is a speedy and cheap way to aggregate data in any desired grouping.

Be careful to see that particularly historical data and all data for current use are defined properly and have the same base. An example of the problem to beware of here is if cases are the routine method of shipment, but case size changed 3 years before from 12 pair to 10 pair per case.

Know the accuracy of your raw data. The forecast can be no more accurate than the data.

Forecasts are not made once or forever. They demand *feedback* to analyze error rates and to correct or fine tune parameters for future forecasts.

Forecasts will always change (beware if they do not!). Do not let a changing forecast move your master production schedule wildly. Establish some limits within which the *forecast* may change, but the master schedule input will not.

Analyze the data before using it to forecast. Look for spurious numbers that far exceed the value of the rest of the data. Throw these points out, or average them if no justifiable explanation can be found for them.

Be careful when dealing in dollars—are they current dollars or constant dollars with a certain year as a base? In general, the use of constant dollars can also point out some very interesting information regarding prices, market growth, etc. Do not mix constant and current dollars.

Remember that most forecasts will be wrong, though hopefully not far off the mark. Management judgment must ultimately be brought to bear on each and every forecast. Use forecasts as a guide for management thinking, not computer-generated absolute answers.

10

Data-Processing and Software Package Considerations

In this chapter I do not intend to go into much detail concerning what hardware is required for a good manufacturing planning and control system. Most of the major mainframe or minicomputer manufacturers can assemble a data-processing (DP) system that will do an adequate, if not superb, job in this area—consult the DP specialists in this field. Instead, this chapter will focus on manufacturing software.

Considerations that Apply to In-House Design of a Manufacturing Planning and Control System

Required Management Information

Before the manufacturing data base is specified, it is important to examine what management information will be required from the manufacturing planning and control system.

From the data in the data base, the manufacturing planning and control system will produce *output* reports. This raises the questions of what information is to be displayed on these reports, to whom they will be issued, and their frequency of issue. Note the difference in terms here. Data are just that—a collection of raw facts, numbers, dates, etc.; data have to be manipulated, selected, and organized to convey *information* accurately, quickly, and clearly. Questions to be answered at this stage are:

1 / What *information* is required to manage efficiently the manufacturing operation?

2 / Who needs this information? Who should be authorized to see which reports?

3 / Why do people need this information? What will they do with it? What management *action* will result from their having it?

4 / When and how frequently is the information needed?

5 / What level of detail is needed by different management levels?

6 / How will the information be conveyed? By hard copy (paper) or CRT video display?

7 / Do we print out all information or only report by exception—that is, when something is wrong and management action is needed?

8 / How will the reports be formatted? What is the optimal way to format them so that they are clear and allow for fast and accurate reading?

Data Considerations

Of paramount importance to the proper operation of any manufacturing planning and control system is the design of the data base with its associated elements. In a system designed "in-house"—by and for your company—the data organization is critical. Time spent on data-base design will have great impact on processing efficiency, speed, and accuracy.

Good data-base design eliminates data redundancy. Having the same data element in two different records requires double the storage space, and presents an opportunity for the two data elements to differ if two seperate functions are responsible for the data maintenance.

The proposed manufacturing planning and control system should be completely flow charted. Data requirements can be assembled based on the following questions:

1 / What data elements are needed to provide the management information required?

2 / Where in the manufacturing process does the data originate?

3 / When are the data needed? Will they be available then?

4 / Who is responsible for what data? This responsibility covers both the maintenance of data accuracy and the update function.

5 / Who should have access to the data? Is this access look up only, or change in addition to look up?

6 / What audit trails should exist for establishing data accuracy?

7 / What additional data might be needed 2, 5, or 10 years from now when the "base system" is running well? What additional software "modules" might management want to add in the future?

8 / What data will we want to preserve for historical records? Where and how will this information or data be kept? On microfilm? On tape?

9 / How do we allow for system crashes, lost data, etc.?

Answering questions such as these in the design stage of the manufacturing planning and control system helps ensure that the data base is well designed, that the anticipated data-storage capacity is large enough, and that the data base is designed so that new records may be added quickly as data requirements change in the years ahead.

In the data-base design and output-information-report design stages, it is critical that representatives of all users be a part of the design process. This should include people from all functions—i.e., Finance, Marketing, etc.—and all levels—i.e., everyone from the CEO to the lowest clerk or shipping dock material handler that may have to operate a part of the system or read and take action based on a system-generated report. The point is that all this work should be done in the *design* stage. Attempting to amend or tack on late ideas or requirements after the design is "frozen" or the system is in use is extremely time consuming and expensive.

Transactional Sequencing

Another consideration of in-house data-base and manufacturing system design is that of transactional sequencing. Flow charts are especially necessary here, along with the all-important *user* knowledge of how things actually *do* and/or *should* operate in the company's manufacturing operations. Proper timing of data accessing, maintenance, and report generation is necessary to ensure accurate information from the manufacturing planning and control system. For instance, a material move ticket from the receiving dock to incoming inspection should not be able to be generated until a receipt for that material is entered on the data base. This data entry by the receiving dock clerk verifies the bill of lading against the purchase order as being correct in quantity description, etc. A flow chart of this process is shown in Fig. 10-1. The inventory stock status report should not be allowed to run until all the day's transactions (in and out) are posted.

System Documentation

The biggest single failing of most in-house designed systems (and, alas, some software packages) is a lack of proper DP system documenta-

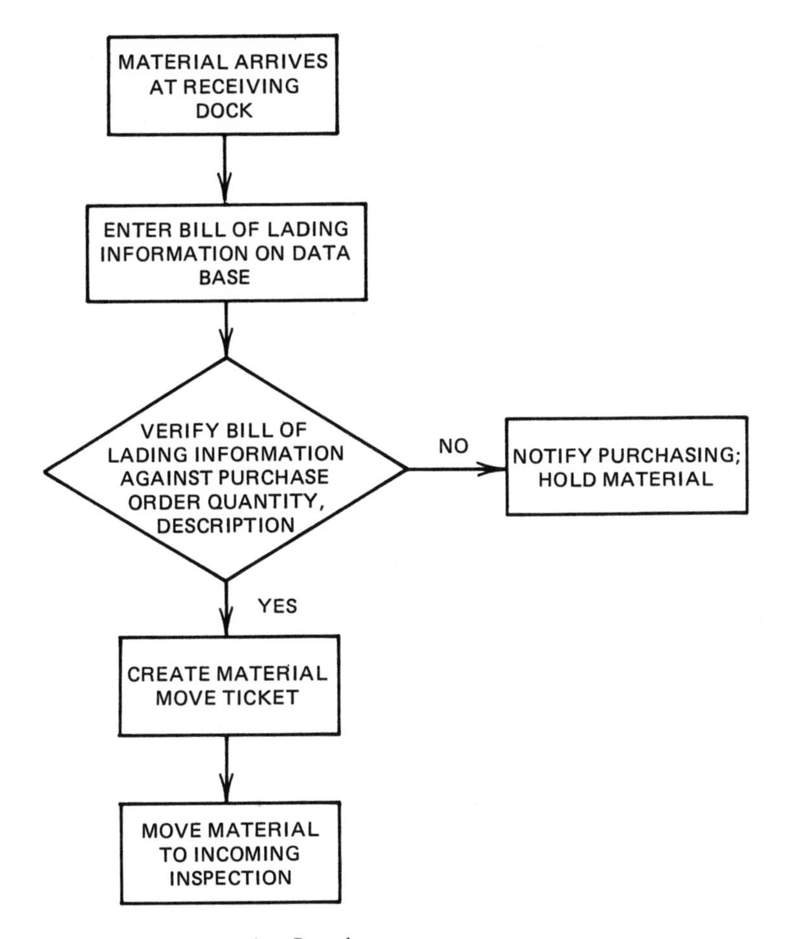

Fig. 10-1. A sample transaction flow chart.

tion. This documentation should include flow charts, the data-base description, job control sequencing, and an easy-to-understand description of system operation. This documentation serves as a basis for in-house system education and for making any changes to the system. All system changes must be equally as well documented with information concerning why they were made, who made them, when they became effective, old and new program listings, old and new flow charts, etc.

System documentation, if done properly, is expensive and time consuming. It also requires a special kind of talent not often available in DP departments—an ability to understand the technical side of the DP operation, but also to be able to communicate effectively this information to people who may know little or nothing about DP systems and concepts.

Many times, in a rush to get a new system into operation, documentation—if done at all—is done hastily and inaccurately, and is weeks or months out of date. It is vital that the DP manager and the system user insist on timely, accurate, and complete system documentation.

A contributing factor to poor system documentation is the high turnover among system analysts and programmers in today's DP world. There is very little chance that the same team that designed and programmed the system initially will be there in 18 months as the system is coming up, or in 3 years when the system may need modification. New programmers and analysts will be employed then, and all they have to rely on for system knowledge is the program listing and any documentation efforts of the past. Do not shortcut the documentation process! If it is shortcutted, it will cause such future problems as added expense, maintenance chaos, and system failure.

In-House Systems Versus Software Packages

Many of the preceding problems can be eliminated by buying established manufacturing planning and control system software packages from either hardware manufacturers or special software houses. These software packages are usually tailored for certain hardware—i.e., IBM or IBM-compatible, Burroughs, Sperry Univac—are written only in certain languages, and are only designed to work with for a certain type of proprietary data base such as IMS, IDMS, or Total.

Software packages offer the following advantages over in-house-designed systems:

1 / There is no point in reinventing the wheel. The amount of time and money necessary to create a good software manufacturing planning and control system package is staggering—millions of dollars for a multimodule MRP system. One reliable estimate from a major mainframe manufacturer is that one line of COBOL code in finished form in a program costs $10 (1979). Some MRP packages have over 500,000 lines of code in them!

Today, while it is true that hardware costs are decreasing rapidly, the opposite is happening to software costs. Programming and system analysis/design costs are increasing exponentially. Most companies simply cannot afford to amass the talent or numbers of people required to design, specify, and code a manufacturing planning and control system. In addition, the high turnover of these people only adds to design expense in the form of lost time for training new people, poor documentation, etc.

2 / Time, therefore money, is saved in implementation. Manufacturing systems software in-house design can be a 1–5 year project, depending on the complexity of the manufacturing environment and the manufacturing control system desired. Why not save this time and effort and just buy a tape from a reputable software supplier? There will still be plenty of tailoring of current systems needed that will take enough time!

3 / The software package you buy is *proven,* or should be. Any vendor of such a system should be able to take you to a number of his or her successful installations so that you can talk to satisfied users and see the system operate.

4 / The software package is, or should be, thoroughly documented, with flow charts, data-base descriptions, and operational descriptions. It should also have a documented history of revisions made by the vendor. In many cases, user groups exist for popular software packages. These national organizations meet periodically across the United States to sort out installation problems and recommend software changes to their vendor.

5 / The software vendor will be available to help your company install its system. They have consultants or experts in all phases of system installation, from DP to manufacturing to education, that can help you when needed. Good software vendors have reputations to maintain, and they do not want installations that have failed for *any* reason on their records. They will work hard to help you make your implementation a success.

Many consulting companies that do not have a manufacturing software package of their own to sell say "your industry or company is special and has requirements that no commercially available system offers. Let us design a manufacturing planning and control system specifically for your company!" While every company likes to believe it or its manufacturing process is unique, this is just not so! Most good manufacturing software packages available today are designed broadly enough so that they can be tailored quickly to any manufacturing environment.

At least one major mainframe computer company is using its manufacturing system software package to sell its mainframe. In other words, they say: "If you want this extremely good manufacturing system, you will have to buy our mainframe on which to run it"—a good sales tactic! This is not bad from a buyer's point also, as you know the system will work efficiently for you. At some point, buying a new clean DP system assures you that you have state-of-the-art equipment *from one source* that is all designed to interface and work well. Today's DP departments all too often contain a hodgepodge of hardware and software accumulated over the years that does not function optimally.

In most situations, a good manufacturing planning and control system software package is far cheaper and faster to install than any in-house design.

The MRP Software Functional Specification and RFP

After the initial "catalog shopping" has been accomplished, it is time to seriously investigate a narrowed field of MRP system software vendors. It is at this point that a request for proposal (RFP) should be created that will be used to solicit proposals from MRP software vendors.

A crucial part of this RFP that is often omitted is the company's functional specification of how its MRP system has to work in its business and the DP environment the MRP system will operate in. This functional specification document concentrates on what any manufacturing planning and control system to be acquired by the company must *do*—in a detailed manner.

The functional specification starts by presenting an overview of the company—its present products, markets, sales methods, manufacturing facilities, and, possibly, its future manufacturing plans.

Second, it details the DP operating environment the new MRP system is to work in. This section must be quite specific in listing all the present hardware, the communications devices, and the data base used by the company. In addition, any interfacing requirements that will have to be satisfied by the new MRP installation, such as linking the current order entry system to the new MRP software, should be outlined.

Finally, the functional specification deals strictly with the manufacturing environment and what functional tasks the desired new MRP system should perform. Rough statistics are first presented—such as the number of SKU's (stock keeping units) in each of the company's stock rooms, how many manufactured versus purchased parts there are, how many levels there are to the average bill of materials, how many operations in the average routing, etc. Then, important tasks the new system must accomplish are detailed, such as the following:

Typical MRP Software Functional Specifications

Master production scheduling

must be capable of supporting two-level master scheduling;

must be a bucketless system.

Material requirements planning

must be a net change system;

must be capable of handling serial number lot effectivity;

must have firm planned order capability.

Capacity planning
must be able to perform resource requirements planning using planning bills.

Creating such a functional specification is an invaluable procedure for any company seeking a new manufacturing planning and control system. It forces its Manufacturing department to *think* about how their company works and, more important, how they want it to work in the future. It also isolates areas where more knowledge is needed—to be obtained either from an outside consultant or from software vendors. Writing a functional specification forces a company to do "up front" the work they would eventually have to do anyway. Not only will including a functional specification in a RFP save you valuable time in working with vendors, but it will serve as a filter to eliminate early in the process vendors who cannot satisfy your company's requirements. Software vendors like to receive a functional specification because it saves them time in qualifying candidates. Furthermore, it conveys a good picture of the MRP knowledge level of their sales prospect.

The RFP should then be mailed to a selected list of software vendors. Besides the functional specification, the RFP should include a projected timetable for software purchase and implementation, the name(s) of your company's personnel to be contacted, guidelines and criteria on how the vendor's presentations are to be evaluated, and other general terms.

Manufacturing Software Package Ratings

A question that often arises is "Which of the 80 or more commercially available manufacturing planning and control system software packages is best for my company?" Here, expert guidance is available. Darryl Landvater and Ollie Wight (Manufacturing Software Inc., Williston, VT) publish MRP software evaluation reports on a great many major software packages. These ratings are excellent. Money spent studying them will be returned many times in the future, if your company is serious about acquiring such a system.

Through your company's RFP, vendors that offer software packages designed for your hardware, programming language, and data base are scheduled for their sales presentations. Once you have worked with these software suppliers and narrowed your field down to two or three vendors's systems, I recommend buying Landvater and Wight's manufacturing software rating reports as an aid in making your final decision.

Time-Sharing Considerations

Another consideration in selecting a manufacturing planning and control system, especially for the small company that has little or no DP hardware investment and limited capital, is to obtain such a system by time-sharing. While expensive for large companies that make many complex products, time-sharing is an excellent way for a small company to get the necessary manufacturing planning and control system in place and operating very quickly. Then, as the company grows it can evaluate the constantly changing hardware scene and buy or lease the computer it needs. Many time-sharing companies are making it easy to switch to an in-house computer by selling you their licensed software that you have installed, and leasing or selling you a minicomputer after a year or two of time-sharing service.

Caveats for Buyers of Manufacturing Planning and Control Systems Software

Some caveats are listed here for buyers of any manufacturing control system software:

1 / What is the total installed cost including software changes (if any), documentation, consulting days, education programs, maintenance package, etc.?

2 / Can the system adapt to your company's growth and changing future needs?

3 / When can the new system be up and in full operation? Who is liable if these dates are not met?

4 / What is the vendor's reputation for integrity, service, education, consulting, and technical help. How fast will they have someone at your plant if their system develops a problem?

5 / Will modifications that your company makes void the vendor's warranty on the system?

6 / Where is the vendor likely to be in 5–10 years? Acquired? Out of business?

In short, buy an established manufacturing planning and control system software package, but buy it wisely, and treat it as the major buying decision it really is. Its cost is high, but the payoff will be higher!

11

The People Side of Computer-Based Manufacturing Planning and Control Systems

Now that you have the technical background and appreciation for what a computer-based manufacturing planning and control system can do for your company, all you have got to do is to buy and install such a system, right? WRONG! *Nothing* is going to happen until you get your company's *management people involved.* You will need to educate them as to how a computer-based manufacturing planning and control system works and the benefits such a system will bring to your company, and get them firmly behind your efforts to install such a system. More wasted years, dollars, and careers can be attributed to not having people—especially the *right* people—ready for these new manufacturing planning and control systems than any failure of the hardware or software aspects of the project.

Top-Down Support from Management

The first requirement for *any* successful manufacturing control system, be it computer-based or not, is to have the complete support and active *commitment* of the CEO and other top level managers. This is the *crucial* factor that so many middle or lower-level managers think they can circumvent or do without. Why does this naive attitude prevail?

Seeking top management approval and support is time consuming. Getting a CEO's *yes* generally means months of studies, presentations, meetings, arguments, etc. Sometimes, lower-level managers *assume* they have a yes when, in fact, they do not!

Seeking top management approval means you have to "go public" with plans, anticipated changes, new ideas, and maybe even new requirements for existing people, or new requirements for new people!

Why Top Management Support is Needed

Why is this top management support needed?

1 / A properly designed and run computer-based manufacturing planning and control system is going to change the entire nature of the manufacturing organization. The implementation of such a system can raise obstacles to change within the present manufacturing organization that have to be overcome. Power in the present organization usually rests in the hands of managers who have been with the company in their present function or position for many years. In most cases, *new systems mean new power centers.* Since people do not like to lose power they have worked hard to acquire, they will be most resistant to any new system that threatens their status quo. With a new manufacturing planning and control system, for instance, the following could happen:

In some companies, the master scheduler will perform the scheduling function that Marketing, Merchandising, or the VP of Manufacturing now performs.

In some organizations, the purchasing manager is currently responsible for inventory control, material handling, etc. With the new manufacturing planning and control system, he or she may no longer have control over these functions.

Manufacturing will be strengthened as a function. For the first time, it will have the information and controls to plan and operate manufacturing effectively. This may be a threat to some managers in companies that have long been dominated by marketing or finance functions.

2 / The new system is going to cost a lot of money! For a complete computer-based manufacturing planning and control system (including hardware), expenditures of one-half to one million dollars are commonplace. These systems, even when acquired and installed piecemeal, are major capital expenditures! As such, their installation requires a great deal of analysis and planning, and a detailed schedule of financial commitment.

3 / The installation of a new manufacturing planning and control system will take from 6 months, for a small module of the overall system, to 2–3 years for implementation of the entire system. Program commit-

ments spanning this time frame must be carefully planned, and need the *continual* support and review of top management.

4 / Such new manufacturing planning and control systems are going to require new skills in new functional areas. If the ability to learn these skills does not exist in the present organization, then new management people must be recruited, and the organizational structure may have to be redesigned. This organizational change process takes time. Only a senior management team that is firmly committed to the new manufacturing system installation can lead these changes.

No change that is this fundamental to the operation of a company can be successfully made without top management support. What many lower-level people are reluctant to do is to secure this support *before* they commit themselves to a manufacturing planning and control system installation.

Selling Top Management on MRP

How is the MRP program sold to the CEO? The following are some suggestions.

1 / Make an *objective* analysis of the state of the manufacturing operation of the company. Is it inefficient, overstaffed, or always in an expedite mode? Are manufacturing costs high or profit margins too low? What is the company's history of inventory levels, inventory turns? How is it functionally organized? What is the education level of its management? To what extent does it rely on data processing? How does Manufacturing coexist with Marketing/Sales, Finance?

2 / Examine the "state-of-the-art" in your business. What's possible? What's the competition in your industry doing?

3 / Develop a proposal for top management to close any gap found with a new manufacturing planning and control system. This proposal should contain:

(a) Broad design specifications—organization charts, functional flow charts, project schedule charts

(b) Required hardware—costs, availablility.

(c) Required software—costs, availability.

(d) Implementation costs. These costs are often overlooked and can easily double the cost of an MRP installation over the software cost alone. Typical MRP system implementation cost factors as a function of average MRP software package cost are:

MRP software (base)	X
Establishing accurate bills of materials and routings	$0.5X$
Establishing inventory accuracy and integrity	$0.5X$
Conducting an education and training program	$0.4X$
Establishing the production planning process	$0.1X$
Outside consultant	$0.1X$
Total implementation cost not including hardware	$2.6X$

These costs are shown in Fig. 11-1 as a percentage of total installation costs.

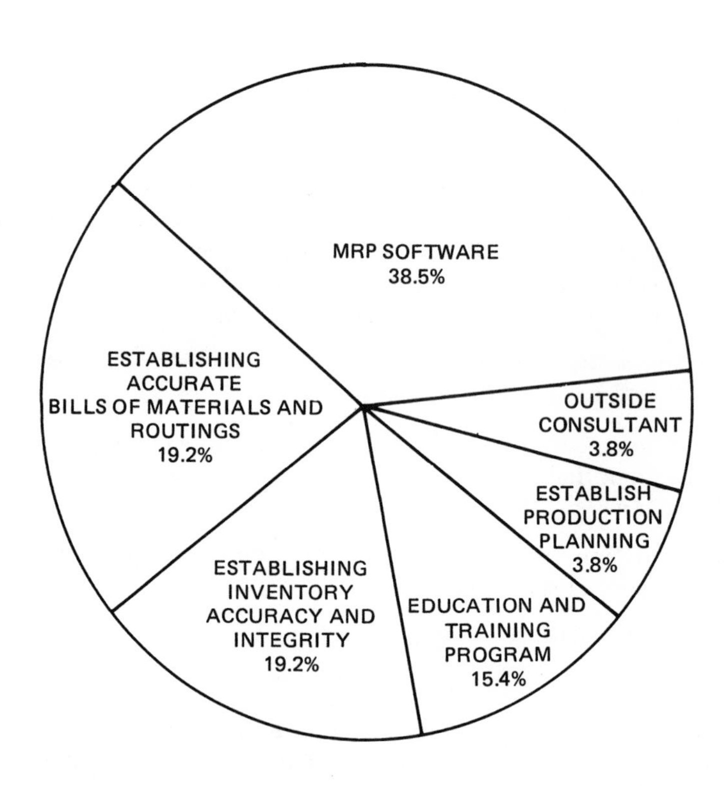

Fig. 11-1. Total MRP system impementation costs.

(e) New functional skills needed to implement and operate the system, estimated manpower savings over the next 5 years.

(f) A pro forma statement of financial changes made possible by the new manufacturing planning and control system—lower manufacturing or purchased goods costs, improvements such as inventory reduction, better customer service, and increased productivity.

(g) An implementation plan—how the installation of each module will be timed, who is to do the work, etc.

(h) Examples for your CEO of other companies that could be visited that have successfully installed a new manufacturing planning and control system.

All this should be assembled in a concise, easy-to-understand, well-illustrated presentation made up of two or three easy-to-digest sessions. The key part of this presentation is that it will be an educational process. You have to *educate* and *sell* the CEO and other top management the fact that your proposed new system is financially rewarding and necessary for the continued growth and success of the company. If the CEO says no, chances are you have not done a good enough *educational* job. But do not stop trying. Keep educating people and refining the proposal until you get a yes.

What all of this means is a constant job of educating people as to the advantage of a computer-based manufacturing planning and control system. Regardless of whether you can aim at the CEO directly, or have to proceed upward a level at a time, education is the key to getting the required top-down support so vital to the success of the new manufacturing planning and control system.

Establishing a Company Education Program for MRP

Once the CEO says yes, too many people put away all the charts, books, and slides and attempt to get on with the job, only to have the project still fail! In truth, once you have the top-down support you need, your education job is just starting. Now you have to educate the majority of the company's personnel!

The key people to educate now are those who will be the users of the new system, and the people who will depend on system outputs. There is no company function that will not be affected in some way by the new manufacturing planning and control system.

The education process should focus on five main items:

Why the company needs this new system.

What the new system will do for the company.

How the new system will affect each user.

Why the user is critical to the system's success.

That the system is not a threat to anyone's job—it represents an opportunity for all people to learn and grow.

Before you install any part of the new system, it is vital to start this education process. The education program should be a multifaceted process that includes:

Involvement in the American Production and Inventory Control Society (APICS). If your company's DP and manufacturing people are not in APICS, get them involved. Pay for and encourage their attendance at local chapter meetings and regional conferences. Encourage them to get APICS certified. Encourage your personnel department to list APICS certification as a plus for incoming new help.

The use of available video cassette education programs on the different aspects of modern manufacturing planning and control systems. Schedule your people for viewing times, and test them on their new manufacturing prowess.

The use of outside consultants and educational programs. Send your people to seminars put on by computer hardware and software vendors, local colleges, and the following professional associations:

AMA—the American Management Association

APICS—American Production and Inventory Control Society

IMMS—the International Materials Management Society

NAPM—the National Association of Purchasing Management

SME—the Society of Manufacturing Engineering

NCS—the Numerical Control Society

The establishment of a company library of manufacturing- and APICS-related literature—pamphlets, and the classic books by Plossl, Wight, Mather, Orlicky, Smith, such as those listed in the Bibliography.

The use of vendor education material. Once you have selected a software vendor, education should then specifically concern how the vendor's particular system works, and how you intend to operate it in *your* manufacturing environment.

The important thing about this education process is that it involves everyone in the company, from CEO to stock room clerks and material handlers—even your suppliers who, after all, will be depending on the system for their requirements. Of course, not everyone needs to learn about the system at the same level of detail. Programs and courses should be carefully structured for differing job functions and management levels.

The education process described above has two essential features:

1 / Establishing and running the education program is a full-time job for someone.

2 / The education and training process should never stop. Even 2–3 years in the future, you will still have to familiarize new personnel with your system's operation, inquire from current employees what changes they would like made in the present system, and familiarize current users concerning changes and improvements you have made to the system.

Graduate Education In Manufacturing Planning and Control Systems

One of the places where education concerning computer-based manufacturing systems has been notably missing is in MBA business school courses. Manufacturing courses are rare in most schools. In many, a brief course in Operations Management is all that is offered, and usually these emphasize "old standards" such as order-point inventory control, quality control, plant layout, and PERT/CPM. These are all good background, but are hardly relevant to what is going on in today's manufacturing world.

One of my aims in writing this book is to see it become the basis for a required course on computer-based manufacturing control systems in most MBA curriculums. I hope that this book and any course in which it is used will then stimulate more people to get involved in manufacturing.

12

Computer-Aided Design

CAD (computer-aided design) and CAM (computer-aided manufacturing) unfortunately do not have standard definitions in industry today. Some people interpret these terms quite narrowly to include only computer-aided drafting and numerically controlled machinery. Others define the term CAD/CAM to encompass *any* application of computers to the design and manufacturing process. In this book, we will take a compromise approach to these definitions of CAD, in this chapter, and CAM, in Chapter 13.

Computer-Aided Design

We will look at CAD as encompassing the following general topics:

- Computer-aided design or engineering
  - Computer-aided drafting
  - Computer-aided analytical testing
- Group technology
- Computer-aided process planning

What Computer-Aided Design Is

In the CAD process, engineers usually work at cathode-ray-tube (CRTs—popularly known as "tubes" or "scopes") work stations (Fig. 12-1) where they create a part by "drawing it" on the CRT screen.

Courtesy of Applicon Inc.

Fig. 12-1. A CAD CRT work station. The operator at the Applicon system's interactive graphics work station creates a drawing by using an electronic pen and tablet and a typewriter-like keyboard, and views his work on a CRT display.

The part's geometric shape can be input to the computer and, hence, "drawn on" the CRT screen using a number of devices such as a standard computer terminal keyboard, electronic stylus, or an electronically sensitive menu tablet of commands that can be selected by the designer. To input part designs from existing prints, a digitizer can be used by the designer to pick up *X-Y* coordinate points that define the part's shape and size.

How CAD is Used

Once the part is entered into the design data base, most modern computer software can display the part from any view by rotating it about any one of its three axes. In addition, a section of the part can be created by passing a cutting plane through any point on the part at whatever angle the engineer wants. An isometric or three-dimensional view of the part can also be selected. Furthermore, the part can be scaled up or down in size automatically, and the viewer can "zoom in on" or "blow up" any portion of the part's image to create or view greater detail. Hidden lines can be shown or deleted from any view. Script or text can be added to the drawing, and the part can be dimensioned automatically in most cases after a minimum of basic dimensional information is input. On assembly drawings, CAD can be used to show places where parts interfere with one another.

A real benefit of using a CAD system is that after the part's drawing is created, it can be stored in the computer's memory (or on disk or tape) and instantly retrieved later for duplication or further modification.

Once the part is in the computer, drawings can be created by using automatic drafting machines. These machines, shown in Fig. 12-2, remove the tedium and toil from the drafting process and can produce extremely accurate prints in minutes instead of hours or days.

Engineering Analysis with CAD

In most CAD systems, a wide variety of analytical software is available for mathematical testing and modeling of part designs. For instance, programs can automatically calculate the volume, surface area, and weight of a part. Engineers can establish finite-element models as shown in Fig. 12-3 of their part designs which they can then expose to test stresses while watching the part deform on the CRT screen. Printouts of stress and strain values at every node can then be obtained from a terminal printer.

Once the part's geometry is defined, its geometry then becomes the basis for future part machining operations and, indeed, even the tool and fixture design for the equipment needed to manufacture the product. For instance, for a part that needs turning on a lathe, the CAD

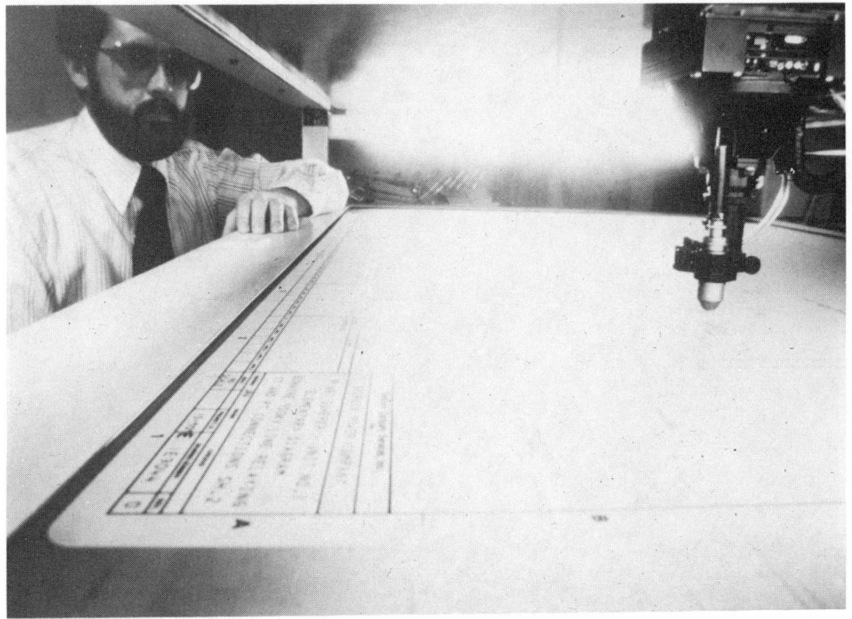

Courtesy of Applicon Inc.

Fig. 12-2. A computer-run drafting plotter.

software defines the cutter location (CL), or tool path, for the part's geometry. This CL becomes the basis for further computer processing to turn it into numerical control (NC) instructions to run specific machine tools.

The Benefits of CAD

The list of benefits from CAD seems endless. Productivity improvements resulting from the adoption of CAD in an electronic or

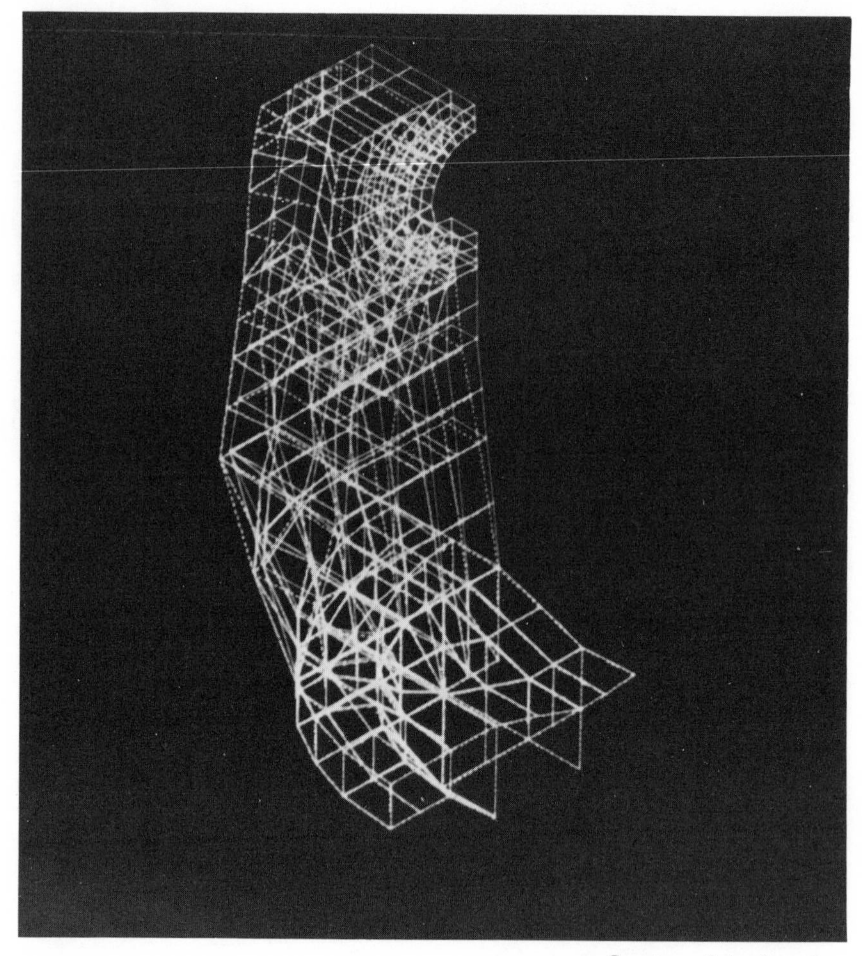

Fig. 12-3. A CAD finite-element model. *Courtesy of Applicon Inc.*

mechanical manufacturing facility range from 3 to 1 to 6 to 1 with isolated cases being even larger! Specific benefits are discussed in the following paragraphs.

1 / With CAD, the design process is quicker because the computer software can be utilized to save time in such steps as drawing lines, calculating dimensions and tolerances, rotating views, sectioning views, checking interferences, etc. In addition, mathematical errors are eliminated, presuming the data are accurate.

2 / Once the part design is in the computer, modifications to it can be made quickly and cheaply. For instance, a simple part can be lengthened 3 inches in just seconds, and the new completely dimensioned drawing

(all views) can be produced just minutes later. It is here, in part modification, that productivity improvements up to 100 to 1 can occur with CAD. Additionally, there is never any worry with CAD that someone is working from an out-of-date print. All references to a given part number will produce the same up-to-date CRT image or drawing for everyone.

3 / One underappreciated advantage of CAD is that its increased speed allows many more part designs to be tested and evaluated before engineers or stylists select the final one. This fact will have a great effect on product quality as firms learn to take advantage of it. Alternatively, for any given design, CAD makes possible a greatly reduced lead time to design and prepare engineering drawings for the part.

4 / In conjunction with the previous point, CAD becomes a viable substitute for the model-building process that used to be the norm for complicated new engineering designs. Previously, such designs were submitted to a model shop where a scaled down model was laboriously (and often slowly) created. Now, CAD allows the "model" to be built on the CRT screen, altered and tested as desired, viewed from literally any direction (including from within!), and printed out in hard copy form as desired.

5 / CAD removes all the drudgery from the drawing process by eliminating the manual drafting process. No longer is erasing the drawing and keeping it clean a worry. No longer do draftsmen spend hours revising a drawing only to have to start over if a mistake is found or someone wants to try the design another way.

6 / CAD enables text in almost any font to be added to the drawing *quickly*. How many painful drafting hours have been spent trying to make the printing or a drawing look neat and legible?

7 / CAD allows the engineer to view the dynamics of a situation. For instance, linkages can be moved through their range of movement on the CRT screen to check for their conformance to a correct path or accurate positioning, and lack of interference with other parts.

8 / CAD can graphically show the results of finite-element analysis. This enables a quicker evaluation of designs to occur.

9 / Modern CAD systems take advantage of color (CRTs) to highlight complex mechanical assemblies or designs for printed circuit boards or solid-state-chip circuits.

10 / CAD enables any view of a part to be selected and drawn correctly quickly. This eliminates the extraordinary amount of time necessary to create alternative views or sections of parts under the old manual drafting system.

11 / CAD software offers a wide range of application programs to users such as automatic nesting programs as shown in Fig. 12-4 for pattern cutting applications (flame cutting of steel, or die cutting of cloth or leather), sheet metal layout, and fixture design.

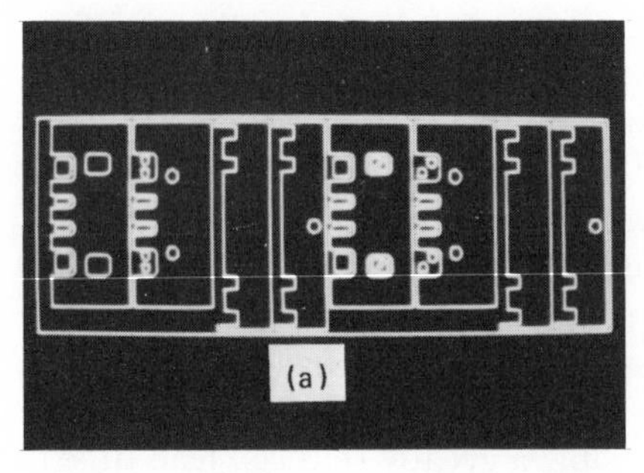

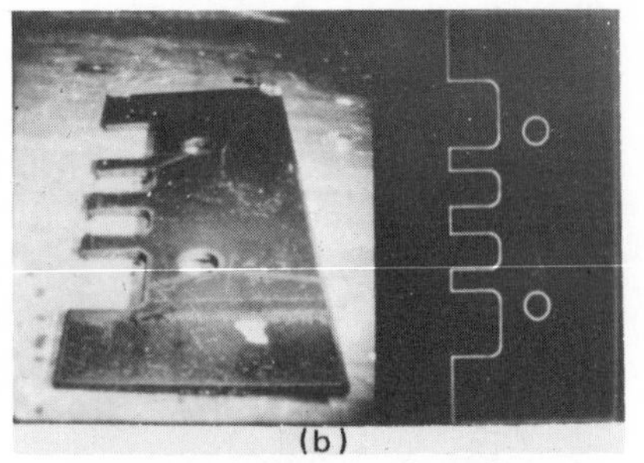

Courtesy of Applicon Inc.

Fig. 12-4. A CAD nesting program layout.

12 / CAD offers hope to companies that have problems recruiting and paying for a large staff of engineers. Engineers are becoming scarcer and more costly every year. In addition, they have very strong ideas about what part of the country and what environment they want to live in. Productivity increases from CAD will mean companies can get along with fewer engineers than in the past.

Today, turnkey four work station color CAD systems complete with typical peripherals such as a digitizer, flat bed plotter, and a CRT hard copy device such as shown in Fig. 12-5 are selling for $400,000 to $500,000. CAD systems continue to improve in capability as new software is developed, computing speed increases, and computer memory cost decreases. CAD has become a must for manufacturers today if they are to be competitive in world markets.

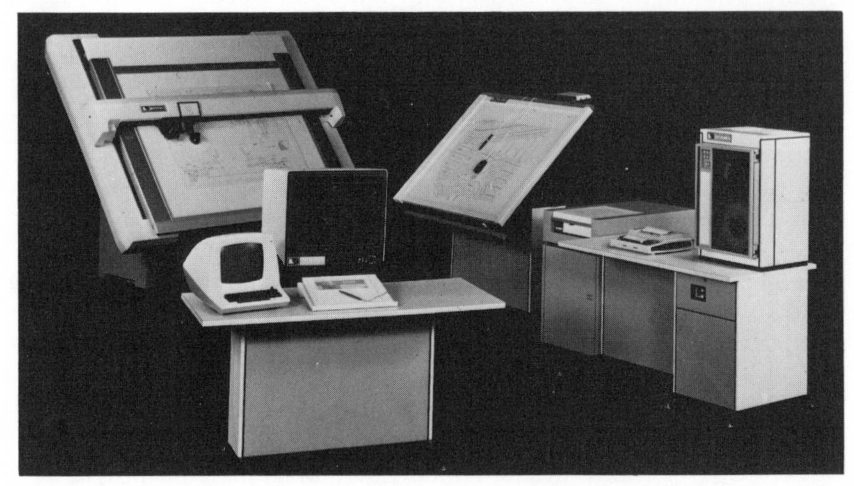

Courtesy of Computervision Corp.

Fig. 12-5. A complete turnkey CAD system. The Designer System by Computervision. In the foreground is the Design Console. In the background from left to right are the Interact IV Digitizer/Plotter, CVD Digitizer, and the Computervision Graphics Processor.

Group Technology

Classification and Coding

In group technology, parts are classified and coded into families of parts according to either similar design or manufacturing process characteristics. Group technology then allows utilization of manufacturing cells where dedicated small manufacturing areas or groups of machines are established to produce parts characterized by their similar design or manufacturing processes.

Group technology's classification and coding is accomplished by using a special coding number that is assigned to each part. This number, separate from the company's assigned part number, is usually from 10 to 30 digits long.

Typical geometric shape and size or other characteristics that can be represented by each coding digit are

Design

fundamental shape—cylinder, tube, cube, etc.
shape characteristics—square end, tapered end, etc.
location of shape characteristics—feature on top, left end, etc.
dimensional groups, i.e., 0–1 inch dia., 1–3 inch dia., 3 inch+ dia.
dimensional ratios
tolerance specification accuracy
weight
volume
material
part function

Process

major process operations—grind, mill, drill, turn
machine tools required for production
lot size groupings: 0–10, 10–30, 30–60, 60–100, etc.
setup time
run time
special fixture requirements

For instance, a 20-digit code number could be used as illustrated in Fig. 12-6. Parts on the corporate data base can then be sorted or accessed by this coded number in addition to the part number or other identifiers.

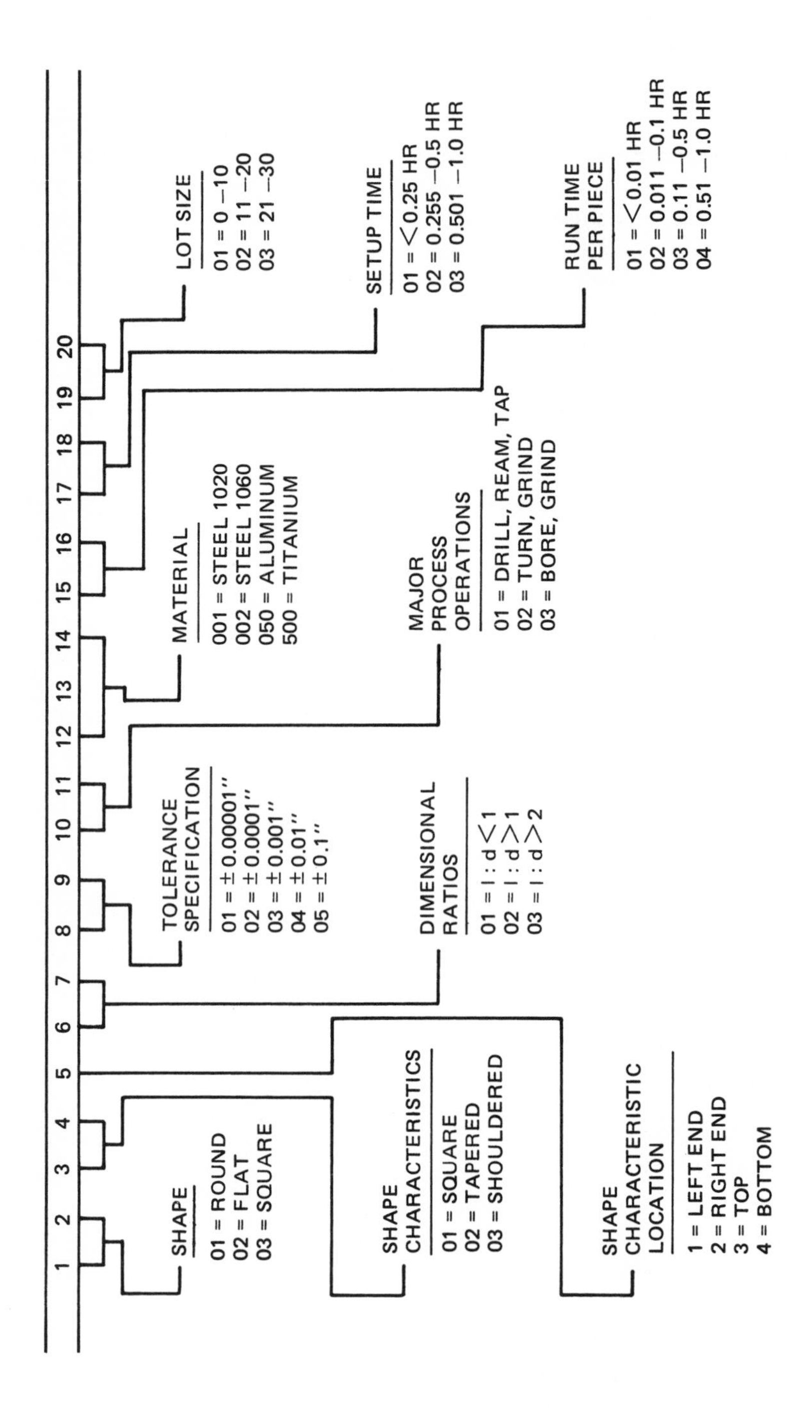

Fig. 12-6. A group technology classification and coding illustration.

Uses and Benefits of Group Technology

Group technology's primary value occurs during the design of a new part. The sketch or conceptual idea of the part is coded according to the company's classification scheme. Then the designer can ask of the data base: "Have we ever manufactured a similar part?" If so, the designer can extract the similar part's drawing and specifications from the design data base for copying and any modifications that may be necessary. Not only is valuable design time saved this way, but unnecessary part duplication is eliminated.

With all the parts in a company's data base coded, parts can be rationalized. It is not unusual for a company to have 100,000 parts on its data base, of which 40,000 may be active. There may only be 5000 of these parts that are really different once group technology principles have been applied.

For standard families of parts, standard routings can be prepared that need only slight modification for a newly designed part. The result is faster, more correct routings at less cost.

Such standard routings can then be used to group production machines in logical "cells" dedicated to the manufacture of one or more families of parts. In addition, similar parts or families of parts requiring approximately the same tooling and machinery can be grouped so that production setup time is reduced, as shown in Fig. 12-7.

There are three cost-related uses for group technology. The first is to review product costing or designs for anomalies. A given family of parts can be graphed on a part versus cost basis as in Fig. 12-8. Anomalies or outliers such as part A will thus be exposed and can be investigated to ascertain if the cost is wrong, or why the part costs so much.

The other side of this cost line—a second and equally valuable use for group technology—is that once a company possesses such cost curves, a cost estimate of any new part design can be derived very easily and quickly. In the example, if we have a $1\frac{1}{2}$ -inch-diameter bushing to design, we could be reasonably confident that it should end up costing approximately $10.00.

Third, with cost tied to the group technology code, one could easily identify, for instance, all active parts made out of aluminum in a company. The annual usage of these parts could then be ascertained, and the effect of a price increase in aluminum could be quickly forecast for the company's pricing or budgeting process. Needless to say, base yearly usage figures could also be established for different materials by this technique.

Based on machine tool processes required, an explosion of next year's

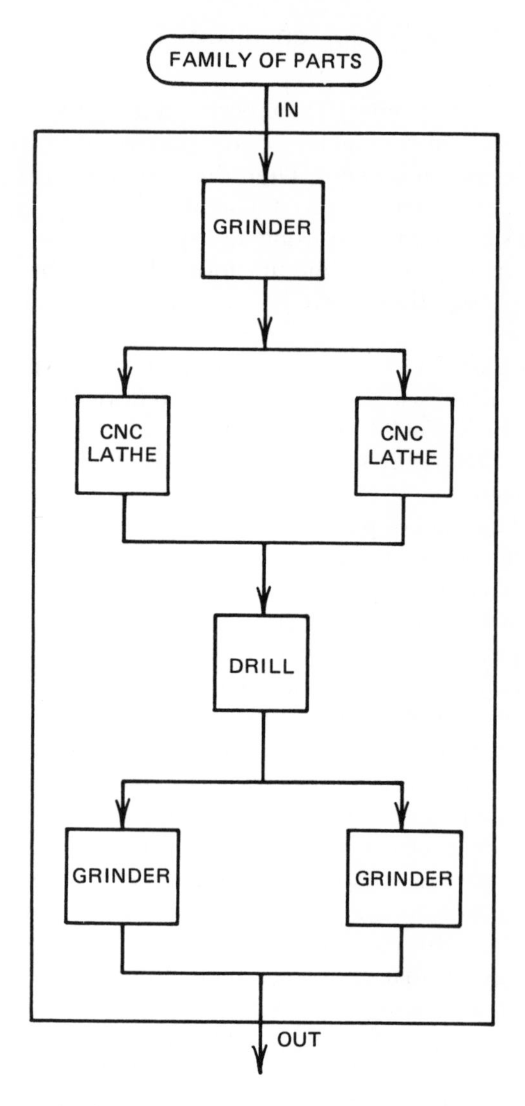

Fig. 12-7. A group technology manufacturing cell.

part requirements through the group technology coding would highlight changes in process requirements that might dictate the need for a new mix of machine tools.

The use of group technology can greatly simplify the production scheduling process by reducing the necessary number of parts, production machines, and routings. In one company, the number of machine tools and routings used to make 324 parts went from 22 and 115 before

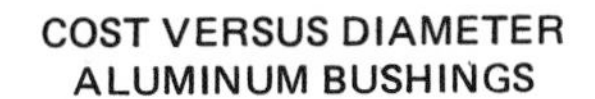

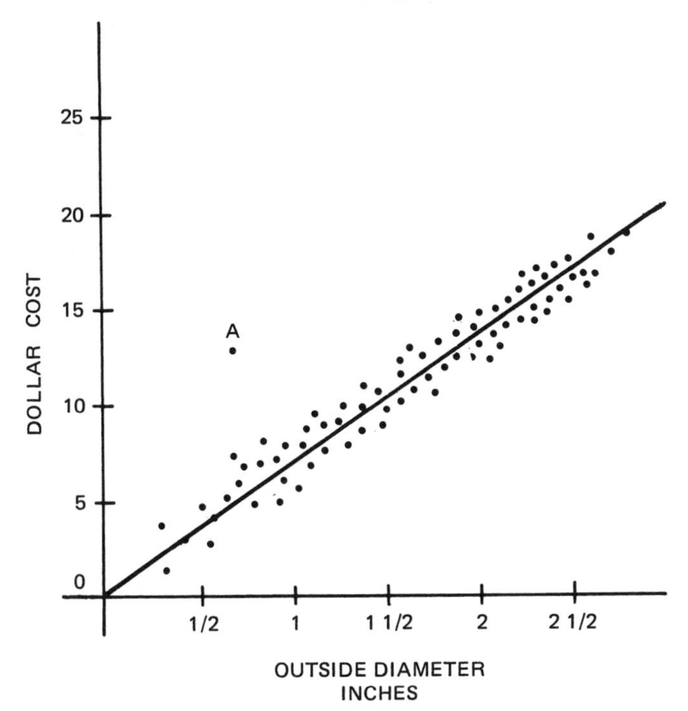

Fig. 12-8. Group technology part cost application.

the use of group technology to 7 and 70 after group technology's adoption.

In another company, the following changes resulted after group technology was adopted:

	Before GT	After GT
No. parts produced	534	534
No. machines	45	4
No. routings	184	10

These are only some of the uses found so far for group technology's classification and coding schemes. The classification techniques used and results obtained from their use are really limited only by the user's imagination.

Computer-Aided Process Planning

Computer-aided process planning refers to the method of preparing routings or process plans using computer assistance.

The Variant Method

Recently, two approaches have evolved for this task. The first and most popular of these is the variant approach where process plans are generated for families of parts that have been classified under the group technology concept.

In the variant computer-aided process planning technique, the process planner calls up existing routings based on a similarly coded family of parts. Then, using a "same as/except for" technique, the family's process plan is quickly amended to cover the individual part in question.

To create a plan from scratch in the variant approach, the process planner can select from a computerized menu of operations based on part characteristics. Here, the process plan is built line by line using company-oriented standard text. This procedure is somewhat similar to a manual process, but the standard text, of course, is stored on the company's data base ready for retrieval, modification, or hard copy duplication.

The Generative Method

The second, far more sophisticated computer-aided process planning technique is the generative method. Here, the computer analyzes the part under consideration and based on part geometry, material, etc., generates a process plan completely automatically. Not only is the sequence of operations generated, but the computer also selects the best-suited company machine tools and calculates a probable machining time for each operation. From this base, it is a simple matter for process planners to review and edit this plan and revise time allowances after the first sample-parts are run.

The generative method is just beginning to be implemented in U.S. manufacturing plants. The key to its more universal adoption will be the development of better three-dimensional solid geometry modeling software. In addition, much work remains to be done on the logic of process planning based on this part geometry.

Benefits of Computer-Aided Process Planning

Comptuter-aided process planning reduces the preparation time of new routings from hours to minutes. Another of its benefits is that process plans are greatly improved owing to standardized operation terminology. Process planning is yet another key CAD function that takes advantage of computerized part design information stored on the company's data base. It forms a vital link in the design and manufacturing process.

13

Computer-Aided Manufacturing

Originally, computer-aided manufacturing (CAM) referred to machine tools run by programs on punched paper tapes. Now the meaning of CAM has expanded, as has the amount of computer applications possible in manufacturing, to include many other subjects. We will consider CAM to include the following general topics:

NC, CNC, DNC machine tools
Robotics
Programmable controllers and microprocessors
Automatic storage and retrieval systems
Flexible manufacturing systems
Computer-aided inspection

NC, CNC, DNC Machine Tools

The first numerically controlled (NC) machine tools were controlled by punched paper tapes that told the machine what to do in a sequence of operations, and where in *X, Y,* and *Z* coordinates to position a cutter or tool. (See Fig. 13-1.)

As was mentioned in Chapter 12, a part's geometry dictates the tool position for the work desired. This cutter location (CL) path can be

Courtesy of Hardinge Brothers Inc.

Fig. 13-1. An NC machine tool.

created by a computer program such as APT or by the software in some CAD systems. Once the CL information is checked for accuracy, it is put through a postprocessor—unique for each type and brand of machine tool—to be converted into a final NC program. If the program is for an NC machine, it will be produced on a paper tape. Each NC machine then has its own supply of paper tape NC programs, such as shown in Fig. 13-2, that generate instructions for making certain parts.

NC machine control can be either of two concepts. *Point-to-point control* moves the tool or the workpiece holder to a desired position and then the desired work is performed. In *continuous path machining,* motion between the tool and workpiece is controlled constantly, so that the complete tool path has to be specified. Continuous path NC is more complex owing to the fact that motion along more than one axis of the machine must be controlled simultaneously.

Over the past decade NC machining has been proven to allow a productivity gain of around 3 to 1 in most applications. Other benefits of NC machining are reduced setup time, better part quality, less scrap and rework, reduced operator attention needed (so that one operator might run two or more NC machines), and less operator skill needed.

Courtesy of Numeridex Inc.

Fig. 13-2. An NC machine program tape.

Computer numerical control (CNC) places a dedicated computer in or alongside the NC machine tool. Then, paper tapes are no longer necessary and machine tool instructions can be created and stored electronically in the computer's memory or on tape cassettes or floppy disks. In modern CNC machines, programs can be created and can be edited or changed if necessary on the machine. CNC eliminates problems with tapes getting oily or torn, and the problem of tape storage. Thus, CNC machines (Fig. 13-3) perform the same functions as NC machines, although their method of receiving operating instructions is different.

Courtesy of LeBlond Machine Tool Company

Fig. 13-3. An CNC machine tool.

Direct numerical control (DNC) offers real-time computer control of more than one NC machine at a time (Fig. 13-4). In the figure, many NC programs are stored in a central computer's memory or tape or disk. Not only can this computer control many NC machines simultaneously, but it can gather *feedback* from each machine as to part production rates and machine status. With DNC, and CNC capability at each NC machine, programs can be downloaded from the DNC computer to the CNC machine tool for running. This saves the amount of memory needed for each machine and means that the program only has to be created and loaded in one machine (the host DNC)—not every CNC that can run the job.

NC machines are becoming increasingly complex and more electronically based than ever before. Machining centers (Fig. 13-5) now exist that can mill, drill, tap, bore, and even more. These centers can have 24 or more tools to select from. The many tasks that these machining centers can perform greatly reduce part setup and handling, and simplify the NC programming job.

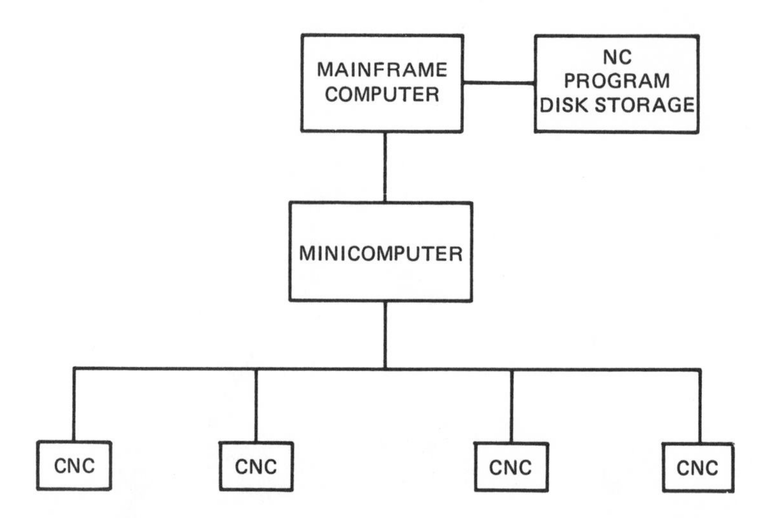

Fig. 13-4. A DNC machine tool hierarchy.

Courtesy of the Monarch Machine Tool Co.

Fig. 13-5. A CNC machining center.

In addition, the application of NC capabilities to other types of machine tools such as grinders, sheet metal forming tools, and gear cutters is generating a large gain in productivity and part quality. The major advantage of numerical control is its exact repeatability of each operation on every machine cycle. Human errors in machine control are inevitable over a long sequence of complex operations. NC lessens this possibility by taking the human control out of the machining process.

Robotics

Robots are computer-controlled devices that automatically perform a programmed sequence of operations. So far in the United States robots have primarily been used in welding and painting applications in automotive assembly plants. Robots are ideal replacements for humans in repetitive, dangerous, or boring factory tasks. Industrial robots are illustrated in Figs. 13-6 and 13-7.

Robots can be programmed by software coding or by use of the "teach mode" to perform tasks. In the "teach mode," a human operator moves the robot through its sequence of operations manually and instructs the robot to remember these movements. Robot software language is now evolving from machine language (assembly code) to higher-level lan-

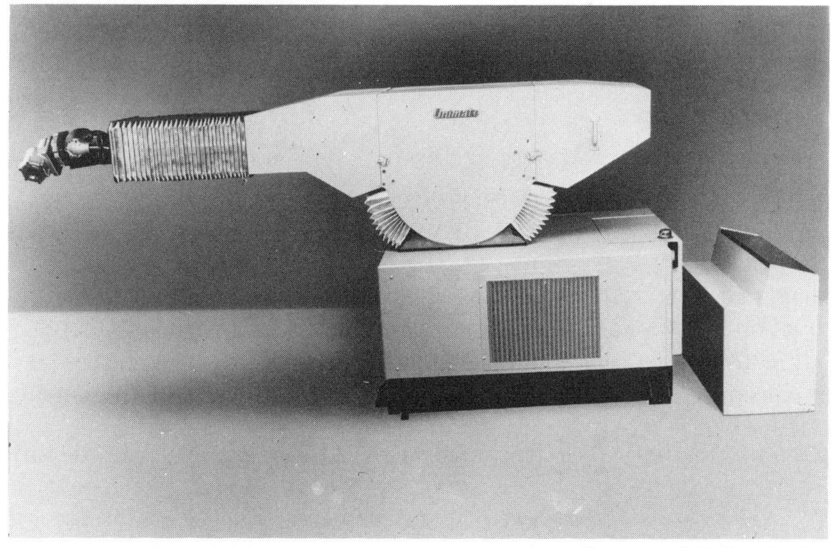

Courtesy of Unimation Inc.

Fig. 13-6. A typical industrial robot.

Courtesy of Cincinnati Milacron, Industrial Robot Division

Fig. 13-7. Another example of an industrial robot.

guages such as BASIC or FORTRAN. This will make the programming of robot operations easier and more efficient.

The following listed and briefly described items are several criteria to be considered when evaluating robots for applications in manufacturing.

Work envelope / The cubic space the robot will be expected to provide arm motion in.

Number of axes of movement / Most sophisticated robots use a form of arm comparable to human's from the shoulder joint to the hand. The more axes of movement a robot has, the more flexible it is in application possibilities. In addition to axes of movement, a related criterion is the type of movement the robot performs—either in spherical coordinates or *x-y-z* rectangular coordinates.

Load capacity / Most robots are designed to work over a limited range of load capacity, such as from 0 to 5 pounds, or 0 to 50 pounds.

Speed / Not only is the speed of robot arm movement important, but often the speed at which the effector (the hand or gripper) acts becomes a consideration.

Type of movement / As in NC machine tools, this can be point-to-point or continuous path movement. The latter requires more computer memory to store movement data points.

Precision / Precision refers to both the concept of accuracy and repeatability. Accuracy refers to the robot's arm positioning capability relative to the programmed target point. Repeatability refers to how closely to a specified location each arm movement comes to the previous cycle's movement. Currently, robots are usable in rather crude applications where exact positioning (i.e., to less than 0.002 inches or 0.050 mm repeatability) is not crucial.

Robot operation can be either based on air pressure, hydraulic pressure, or electric stepper motors.

The technological advances needed to bring about the widespread adoption of robots in factories are the two feedback senses of touch and sight. As pattern recognition software and sensory equipment and adaptive feedback mechanisms become available at even lower costs, robots will be qualified to perform many assembly and part selection tasks now requiring these exclusively human feedback factors. This progress will be welcomed in the United States, as well as in most developed countries, where the supply of people qualified and willing to work in factory environments is steadily decreasing.

Programmable Controllers and Microprocessors

Programmable controllers (PCs) are microprocessor-based devices that can control many processors on machines or in factory systems. Unlike old fashioned hardwired relay-based process controllers, PCs are relatively inexpensive solid-state devices that can be reprogrammed quickly to perform new tasks. Better still, different task sequences or programs can be stored in the PCs memory or on tape so that process control changes can be effected with the change of a tape cassette or with a typed-in command. PCs such as the one shown in Fig. 13-8 are used to control a sequence of factory processes such as for a heat-treat oven—open door, turn on conveyor to move in new parts, close door, turn on

Courtesy of Texas Instruments Inc., Industrial Controls Division

Fig. 13-8. A typical programmable controller.

heat, heat to 1800°F for 6 hours, cool to 800°F for 2 hours, cool to 200°F, signify parts ready for exit from treat heat, open door, move parts out of oven, signal oven ready for new load, etc.

The ultimate advantage of PCs and microprocessors is that any number of them can be controlled by central computers. These devices can then be used to report back on factory floor/process status, and can be reprogrammed quickly for different jobs when necessary.

Automatic Storage/Retrieval System

Automatic storage and retrieval systems (AS/RS) are computer-operated part pickers and stockers. Generally the large-pallet-sized systems for warehouses are called AS/RS, while smaller stock room bin-sized systems are labeled automatic part pickers/stockers. Working either from direct computer-generated and -linked pick or stock instructions, or from manually generated pick or stock instructions, these machines either will deliver a batch of parts to an open random location while recording the location for future reference or will stock a given part in a preprogrammed location or will go to a selected location to pick a part. In warehouses, these AS/RS systems, as shown in Fig. 13-9, often work in environments where pallet racks extend 40 to 50 feet high and over 2000 part stockings or retrievals occur per day.

In part picking or warehouse situations, these systems minimize storage space requirements, and if programmed and used properly, minimize stock access time (Fig. 13-10). In finished goods warehouse situations, AS/RS systems can be tied to automatic part weight inspection to check for correct part quantities. Interestingly, if such systems can deliver parts to a common staging point, how far are we from seeing completely automatic parts delivery to various points on an assembly line?

Courtesy of Eaton Kenway, a subsidiary of Eaton Corp.

Fig. 13-9. An automatic storage and retrieval system.

Courtesy of Lyon Metal Products Inc.

Fig. 13-10. A computerized part picker/stocker.

Flexible Manufacturing Systems

Flexible manufacturing system (FMSs) embody several concepts we have discussed in one computer-run system that is responsible for the complete production of some families of parts. The FMS is a dedicated machining facility and/or assembly line that utilizes a family of parts concept obtained from a successful application of group technology principles. Flexible manufacturing systems, such as shown in Fig. 13-11, encompass raw material storage, part picking, part transportation, DNC machining at several machine centers or work stations, and automatic delivery of the finished part to a finished goods warehouse.

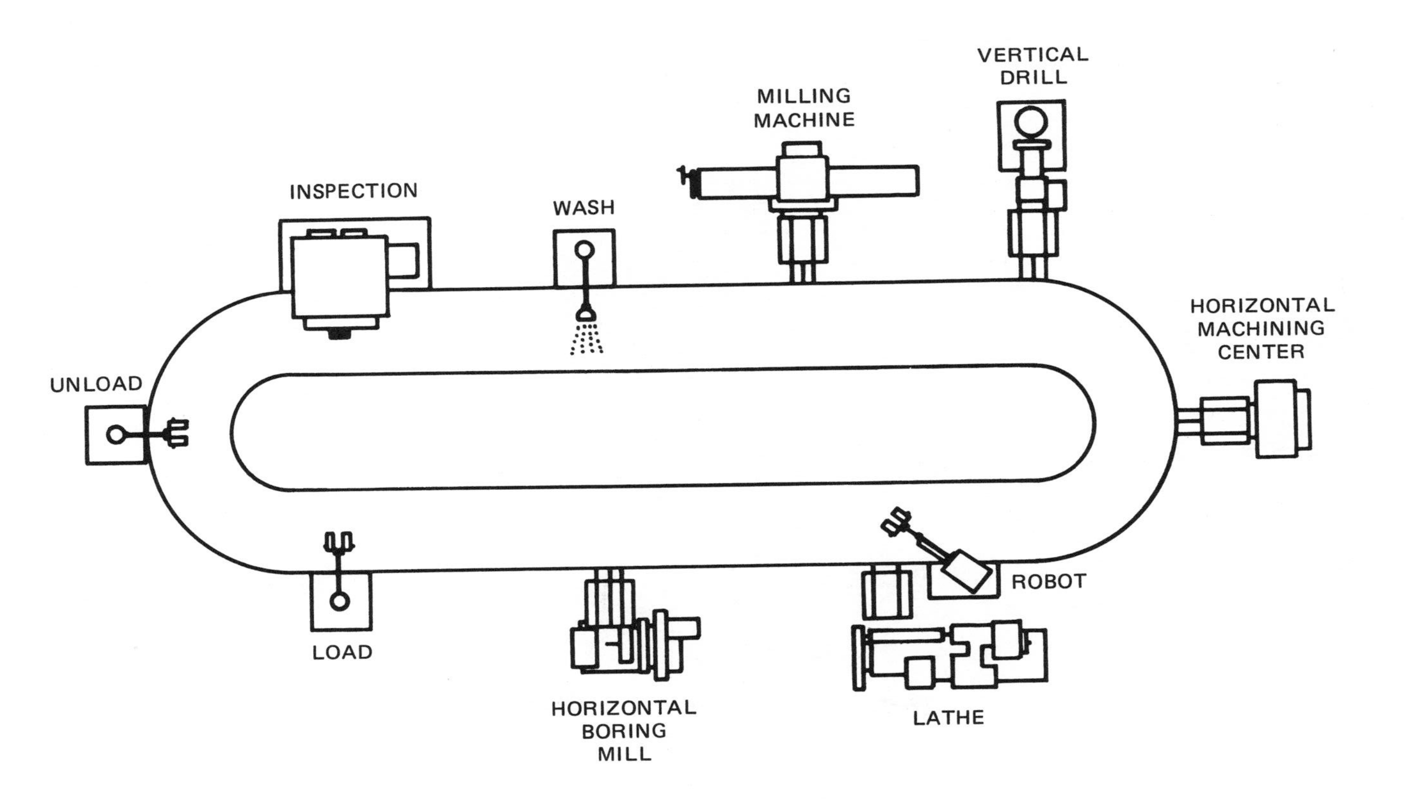

Fig. 13-11. A typical flexible manufacturing system.

The central FMS computer schedules and tracks all production and material movement in the FMS center. Based on a family of similar parts, a FMS can be reprogrammed quickly through downloaded instructions from a central computer to individual machines, conveyors, part picking, etc., to perform a new set of tasks. The major benefit of a flexible manufacturing system is flexibility of manufacturing resource assignment, along with computer-controlled operation.

Computer-Aided Inspection

Computer-aided inspection (CAI) is yet another new use of computerized engineering design data in manufacturing, the time in the quality control area.

In CAI, coordinate measuring machines (CMMs), controlled by software that draws data from the company's engineering design data base, automatically measure parts to see that they have been manufactured to design tolerances specified on the part's drawing. These machines have a probe that automatically moves to a programmed point, takes a measurement, and displays or records the result. The computer can also print out both the required and actual measurement or dimension and/or the deviation, if significant, for the part under consideration. More importantly, the CMM also enables these measurements to be stored automatically in a data base to be used as a basis for further statistical analysis on the part, or perhaps as an early indicator of machine tool wear. These data also serve to record permanently for the company the data necessary to show the part was manufactured accurately, for protection in the event that a product liability suit arises at some future point.

The use of CMM, such as illustrated in Fig. 13-12, in place of more traditional quality control methods typically results in a productivity gain of 3 or 4 to 1.

The Effects of CAM

It is important to note one central feature of all CAM applications we have touched on: This is that with CAM, the skill requirement for an operation is transferred from the operator to the programmer who creates all the machine instructions. No longer must the person running an NC lathe be a skilled machinist; instead the person can be a less expensive and more readily available machine operator. The same trend is evident in all applications of the computer to manufacturing processes.

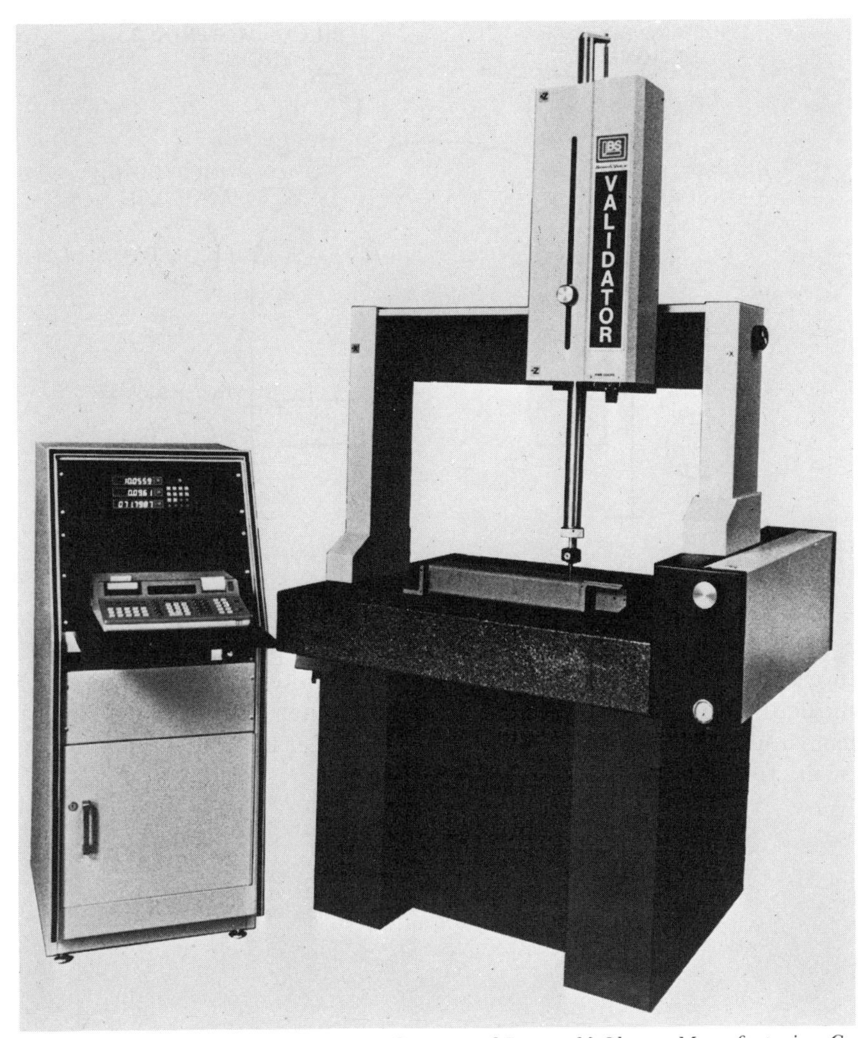

Courtesy of Brown & Sharpe Manufacturing Co.

Fig. 13-12. A computerized coordinate measuring system.

On the other hand, CAD/CAM is promoting the integration of design and manufacturing. These two previously disparate groups under CAD/CAM work from the same data base almost concurrently in the design process to design the part, the tools and fixtures needed to manufacture the part, the bill of materials for the part, and the process plan for the manufacturing of the part. In fact, one progressive company has created cells of people who work as a team through one to three CRTs to carry out the design and manufacturing engineering process. A typical cell is illustrated in Fig. 13-13.

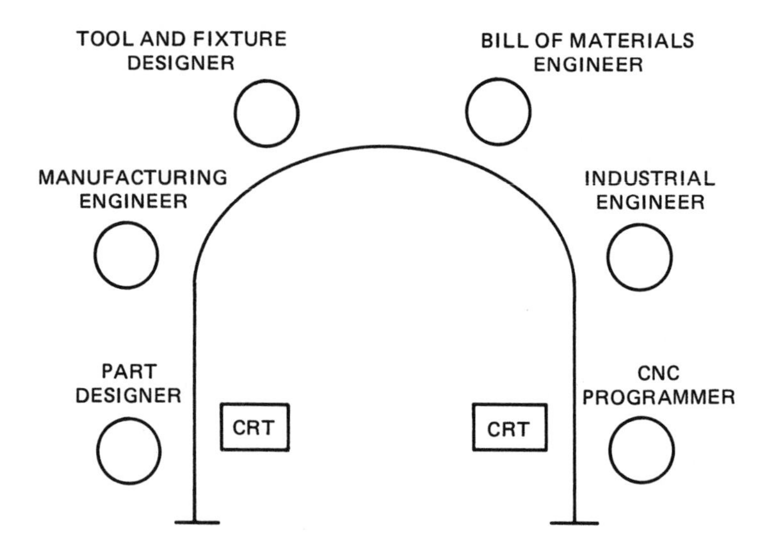

Fig. 13-13. A design and manufacturing engineering cell.

Each cell is responsible for a product family or product group. People in each team gain synergy from their continuous interaction, and not incidently come to have considerable pride in their group efforts. Such a team approach to manufacuring design and operations is surely the way of the future.

14

Factory of the Future

It is increasingly clear that the extent of computer applications in manufacturing and engineering operations will be a major determinant of a company's future success as measured by financial strength and market dominance. Let us now step back and observe the overall trend in these computer applications, and build a conceptual frame of reference as a guide to understanding the Factory of the Future.

The Integration of CAD, CAM, and MRP: Factory of the Future

Cheaper, more-powerful computers, along with the development of data bases, distributed processing, new communications techniques, and new applications software are making possible the integration of CAD, CAM, and MRP into the Factory of the Future. The conceptual framework for this integration is shown in Fig. 14-1.

Some firms already have computer applications that can operate as a package in any two of these three areas in isolated functions; mrp-generated order pick lists, fed directly to an AS/RS part picker's minicomputer, would be one example of this partial integration.

The major task retarding progress in the total integration of these three areas is that of applications software design and coding. Such software is the unifying basis for the Factory of the Future. Today the hardware and communications capability exists for a Factory of the Future, as does the technical expertise. However, we are a long way from having software that can work effectively to unify CAD, CAM, and MRP

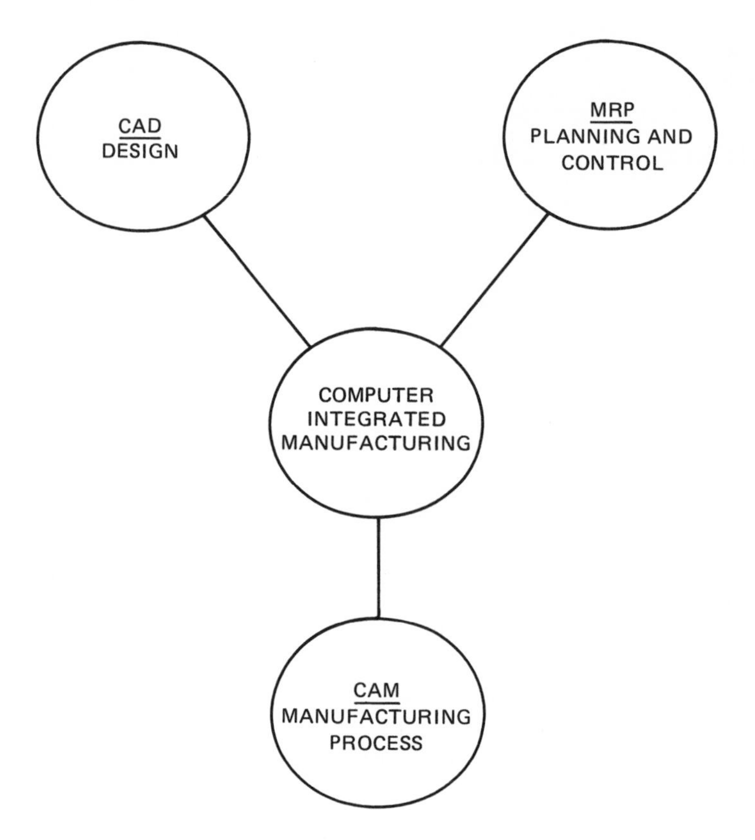

Fig. 14-1. Factory of the Future conceptual illustration.

in a complex factory environment. A significant effort is also necessary to construct the large corporate data bases that properly structure and store the data needed for a Factory of the Future.

Where is Factory of the Future Now?

There are many progressive U.S. and international corporations that are making substantial progress toward a Factory of the Future. The job is immense—in terms of complexity, resources required (manpower, dollars, and time), and education and training tasks. Last but not least is the job of overcoming the human tendency to keep the familiar status quo.

1 / Corporate progress toward a Factory of the Future is being aided by daily advances in technology. State-of-the-art manufacturing software (MRP) is now available on minicomputers that can work in a distributed

processing environment. There are major improvements being made in computer graphics such as color raster displays that provide good picture resolution at increasingly cheaper prices. Robots soon will have limited vision capability and a sense of touch.

2 / While software continues to be a major bottleneck to implementing a Factory of the Future, there have been encouraging recent developments in the area. First among these is the emergence of a highly competitive *software* industry that is providing an improved software applications package to a growing market. There are now over 90 manufacturing software packages available. Independent CAD software is now emerging that frees users from turnkey CAD systems and allows them more freedom in designing their own CAD system and including the necessary linkage to their existing manufacturing data base. Robotics is presently in its infancy as an industry. Major U.S. corporations such as GE, IBM, and Texas Instruments are known to be looking seriously at entering this market.

3 / The challenge from foreign competitors and our recent concern over the United States's declining rate of productivity increase continue to act as major catalysts to spur U.S. industry toward a Factory of the Future.

4 / Many leading corporations are building their own internal capability in a Factory of the Future. General Electric, a leader in the United States in CAD/CAM implementation, has just acquired Calma—a major U.S. CAD company. Black & Decker is heavily involved in its own MRP and CAM development. Boeing and Lockheed are leaders in CAD/CAM capability. Texas Instruments already uses robots of its own design to build and test some of its products. Such companies are moving steadily toward the full integration of all three aspects of a Factory of the Future—CAD, CAM, and MRP. The first step is to integrate these areas internally within the corporation.

The next step though is even more exotic. Major corporations are beginning to visualize the market for a Factory of the Future. Who will be the first to market turnkey Factories of the Future? Who will be the first to sell the software, computers, and production machinery necessary to run a Factory of the Future? The pieces all exist today. The challenge is to turn today's internal tool into tomorrow's market-oriented product.

Implementation: U.S. Government Help

The task of integrating CAD, CAM, and MRP is such a huge project that very few companies, if any, have the resources. The U.S. government, in recognition of this dilemma, is attempting to focus private sector efforts and promote the Factory of the Future concept through its Air Force sponsored Integrated Computer-Aided Manufacturing (ICAM) program. This 7 year $100+ million program is using large U.S. manufacturing companies, such as GE, Boeing, Lockheed, General Dynamics, along with universities, consultants, and software houses, to design and create an aerospace-focused Factory of the Future. One of the government's intentions is for ICAM to provide seed money to U.S. industry as a stimulus to undertake the research and software development needed to tie CAD, CAM, and MRP together.

The federal government is making parallel efforts in several other areas to stimulate U.S. industry to move toward a Factory of the Future. The U.S. Department of Commerce has established a new office for Productivity, Technology, and Innovation that will coordinate two new programs designed to spur innovation and enhance productivity in American industry.

The first of these is the new Center for the Utilization of Federal Technology (CUFT). CUFT will help U.S. industry by transferring new technologies from government to private industry and also between industry sectors. The goal is to increase U.S. industries's competitive position in world markets.

The second program is concerned with the establishment of Cooperative Generic Technology (COGENT) Centers throughout the country. These centers—research-oriented—will target their programs toward helping specific U.S. industries adopt new generic technologies to enhance their competitive positions.

Both of these programs are designed to encourage technological innovation and the *application* of new technology to U.S. industrial processes.

Recent Department of Commerce changes have also made it easier to gain access to new patents through the National Technical Information Service (NTIS). NTIS will also provide incentive through exclusive licensing rights to encourage the development and commercial use of these government patents in high-technology areas.

Recently, the federal government announced that through the Air Force's Manufacturing Technology (ManTech) Modernization program that it would give special consideration to bidders for Air Force Systems Division work who use innovative manufacturing methods as a way to lower costs, improve product quality, and shorten procurement lead times. The Air Force will also provide financial incentives to these contractors so that they can invest in the most superior technology available. This new program will provide a major stimulus for companies to press on with their efforts to develop their own Factory of the Future.

Implementation: How U.S. Business Can Meet the Challenge

There are three major investment areas important to the development of Factory of the Future:

1 / Computer-controlled manufacturing equipment. Obviously, a Factory of the Future will require a huge investment in computer hardware, software, and computer-operated equipment. Fortunately, computer hardware costs are still decreasing as we learn to pack more information into less space. Computer-operated equipment costs seem to rise with the general industrial price level. Software development—based on human logic—will be the costly burden to bear.

2 / Organizational structure and staffing. A Factory of the Future will require a new organizational structure and staffing mix as business operations continue to become more DP dependent.

The field of computer-based manufacturing is growing explosively. New technologies and concepts of manufacturing and manufacturing control arrive daily. Information on these new areas appears to multiply exponentially. To miss any of this new information is to fall back instead of (at minimum) staying even with the pace of progress in this area.

For large multidivision U.S. corporations, at least four new positions might be created to keep pace with this explosive information growth and as a step toward a Factory of the Future. One of these four staff positions would be a Vice President in charge of Manufacturing Systems Development, who is *not* concerned with getting today's products out the door. His or her sole job would be to guide and direct the corporation toward a Factory of the Future.

Under this VP would exist one Manager to correspond with each area shown in Fig. 14-1, i.e., computer-aided design, manufacturing planning and control systems, and computer-aided manufacturing.

These people and their staffs would need active support and much authority from their CEO and the company's board of directors to press on toward a Factory of the Future. In smaller companies, one senior person might perform all four jobs. The critical point is that a knowledgeable and capacity person be charged with the full-time responsibility of guiding the company toward a Factory of the Future.

3 / Continuing education. Today, change occurs at a constantly increasing pace. Technical people become obsolete in 3–5 years. A continuing company commitment to *people*—their education and job satisfaction—will be a vital part of creating and operating the Factory of the Future.

This commitment consists of

Continuing technical or management education for all employees, including top management. While the level of detail between educating technicians and top management should be different, the education experience and exposure to new and more productive ways of doing things is just as valuable to men and women from all management levels.

Creating several alternative career paths for employees that would offer equal career reward for different types of work and personal ambitions.

Investing in community-based vocational education and adult education programs. Such programs would attract and train employees right out of high school, as well as "recycle" older employees whose jobs have been eliminated or changed due to technological innovation.

Conclusions

Today, there is great concern over U.S. manufacturing technology and productivity. There is an emerging movement for the reindustrialization of America in order to make the United States once again the technological and industrial leader of the world. The Factory of the Future offers the way for America to accomplish these goals.

Currently, we sell in world markets. Our industrial competition is global in nature. But as a society, we lack the image of ourselves as a global competitor, and that constrained view is reflected in our declining industrial institutions.

We must take a fresh look at our manufacturing operations—not only to be productive, but to be productive while being effective—doing the *right* things *efficiently*. The Factory of the Future can provide us with the tools to manage the production process effectively.

The Factory of the Future and business of the future will be with us soon. It is clear that progressive companies who expect to be tomorrow's leaders are started down the experience curve toward such a concept. The future will belong, as always, to those who have the vision and courage to invest in it.

Glossary

ABC classification / A method for grouping inventory by decreasing order of annual dollar usage according to Pareto's law. Typically, the inventory is divided into three classes—A, B, and C—where approximately 20% of the items account for 80% of the annual dollar volume (A items), 50% of the items account for 5% of the annual dollar volume (C items), and all other parts are B items.

adaptive smoothing / A technique used in exponential smoothing forecasting where the alpha smoothing constant is periodically revised based on error analysis of previous forecasts.

algorithm / An ordered logical procedure used to solve a problem.

allocated material / Material reserved for a specific order that has been released but whose parts are not yet picked. Once allocated, these parts are not counted as available for future use.

alternate route / An alternative or optional sequence of operations on a routing.

anticipated inventories / Inventories in excess of a base level that are built up to cover extraordinary requirements such as special sales, future (seasonal) orders, plant shut downs, etc.

APT (Automatically Programmed Tools) / An NC programming language.

assembly / A unit of parts and subassemblies joined to make a product or larger subassembly for a higher-level item.

audit trail / A clear sequence of transactions that can be used to document how a number (such as an inventory balance) was calculated.

automatic rescheduling / Where, instead of a planner reacting to reschedule messages concerning the need to reschedule open shop orders, the computer automatically changes due dates on orders where requirements and scheduled receipt dates do not agree.

available to promise / That portion of inventory or a time period's production that is still available or for sale: (beginning available inventory + scheduled receipts) − (customer orders) = available to promise.

axis / A reference line in a coordinate system; also the direction or path of tool workpiece movement.

backlog / The quantity of outstanding orders received but not shipped.

backorder / An unfilled order created by a stockout that is listed as a requirement against future inventory.

backward scheduling / Scheduling backward from the order due date by the time each operation takes, in order to arrive at an order start date.

balanced loading / Loading a sequence of operations where the capacity of the entire assembly or production line is considered and no one operation's work center is significantly over- or underloaded compared to the line's capacity.

BASIC (Beginner's All Purpose Symbolic Instruction Code) / A high-level interactive computer language developed at Dartmouth College.

batch processing / The processing of computer transactions at one time or all at once, not interactive and on-line.

baud rate / The number of bits per second of data transmitted between computers and I/O devices such as CRTs or terminals.

bill of lading / The contractual paperwork that accompanies items shipped by common carrier. It gives a description and quantity of goods, and lists terms and conditions of shipment.

bill of materials / A listing of all parts that make up a product. This listing can show materials, parts, or subassemblies.

bin location / The specific address of the location where a part is stored.

bit / A binary digit, either 0 or 1.

blanket order / A large order that covers a long period of time (such as a year) against which smaller periodic releases are specified. Usually used to secure a firm price or guaranteed delivery of parts or materials by reservation of the vendor's capacity.

bottleneck / The limiting operation in a process, i.e., that operation with the lowest flow rate.

bucketed mrp system / Where all time-phased data occur in aggregated time units called buckets or time buckets. Most commonly, days are grouped into weeks.

bucketless mrp systems / Where time-phased data occur specifically in daily shop calendar dates that are not aggregated into larger time units.

buffer stock / Inventory meant to decouple two operations so that the second operation's production rate is not dependent on the output rate of the first operation.

buffer/storage / Computer memory used to decouple computer processing time from peripheral operation time.

byte / A group of bits of binary computer data. There are usually eight bits to a byte. Computer *words* can be made of several bytes, usually 1, 2, or 4.

CAD (computer-aided design) / The use of computer equipment to design and/or draw a part. The design specifications of this part are then stored in an engineering data base.

CAM (computer-aided manufacturing) / The use of computers to control any number of different machines used in the manufacturing process or in material storage/movement.

capacity / The maximum rate at which a task can be performed, in the case of throughput. Or, how many units one can produce with fixed resources.

capacity requirements planning / A short- to midterm (1 to 12 months) planning exercise based on master schedule/material requirements. These requirements are extended by the needed part's routing times to arrive at labor hours needed to sustain the requirements in each work center; includes not only planned requirements but load due to open shop orders.

carrying cost / The cost of having inventory. Includes cost of capital, space costs, insurance, taxes, material handling, and damaged or lost inventory. Normally expressed as a percentage of unit cost.

central processing unit (CPU) / The main unit of a computer that contains core memory and electronics that are responsible for the control logic and arithmetic processing functions.

character / A letter, symbol, or number, one space long.

closed loop system / A control circuit or system containing a feedback loop or mechanism that feeds back real-time information to serve as a reference for the system's output.

CNC (computerized numerical control) / Numerical control where the machine instructions are electronic (as opposed to NC tape) and are stored in a minicomputer that is part of the CNC machine.

COBOL (Common Business Oriented Language) / A popular business-oriented high-level computer language.

code / Popularly refers to line(s) of instruction in a computer program.

coding / The programming of a set of instructions for the computer.

compiler / A translation machine that converts a high-level program in source language (such as COBOL or FORTRAN) to a machine-language program in object (ASSEMBLY) code.

component / A part or subassembly that is part of a higher-level assembly.

computer / An electronic device that manipulates and processes data according to programmed instructions and reports information as a result of these operations. A basic computer consists of input/output devices, a central processor unit, and memory devices.

computer-aided process planning (CAPP) / Process or routing planning that takes advantage of standard sequences of manufacturing operations stored in a computer. These standard sequences usually are developed for families of parts coded under the group technology concept.

configuration / The way a product is structured or designed to go together.

constraint / A limitation or boundary placed on a given set of circumstances. Usually in optimization problems, such as linear programming, where constraints bound the objective function.

continuous path operation / Refers to machine tool operation where the motion of the tool is controlled totally throughout the tool's movement, as opposed to point-to-point control; involves simultaneous movement on two or more axes.

contouring / The shaping that can be done to a surface by continuous path operation; involves simultaneous movement on two or more axes.

core memory / The main memory of a computer, hardwired into the computer.

cost center / Any department or work group where costs are gathered for accounting purposes.

cost of capital / The cost of money to a firm, usually containing a risk-free portion and a risk-influenced portion.

CPU / The central processor unit of a computer that obtains, decodes, and executes program instructions and maintains status of results.

CRT (cathode ray tube) / An electronic picture tube with a screen on which information can be displayed.

critical ratio / A scheduling priority assignment algorithm that calculates a priority number by dividing the time to the job's due date by the standard time left needed to finish the job, i.e., time remaining/work remaining. Jobs with critical ratios under 1 need expediting or rescheduling. Jobs with critical ratios over 1 might be de-expedited.

cumulative manufacturing lead time / The total lead time necessary to manufacture an item from order to the part being available for use—includes purchased part procurement—measured by summing the longest lead time items needed at every level of the bill of materials that make up the end item.

customer order / A sales order from a customer for certain of the company's standard or custom parts or products.

customer service / Satisfying the customer by delivering the agreed upon product at the agreed upon price at the right level of quality and especially at the agreed upon *time*.

cutter path / The route followed by the cutting tool in order to arrive at the part's desired shape or design.

cycle counting / A procedure for randomly counting inventory at least once per year that is used not only to verify inventory accuracy but to expose transaction and human error patterns so that they can be corrected.

cycle time / The time required to complete an entire sequence of events and to be ready to start the sequence over again.

data base / Refers to data stored in a defined manner (hierarchical, network, or relational) in a computer memory device—usually disks. Under the data-base concept, application programs are not affected by changes in data-base elements.

data file / Similar to data base, but data in a file are usually related to one particular functional application, such as the accounts payable file.

day length / The number of hours in a standard work day.

dedicated computer / A computer used for one specific function only, i.e., shop floor control, NC machine control.

demand / Requirements for a part or product, *not* equal to sales. Demand usually exceeds sales due to stockouts.

dependent demand / Demand that is a direct function of the end or parent item as dictated by the bill of materials or product structure.

direct labor / Labor actually contributed to the production of a part or product, as distinguished from support (indirect) labor; the labor shown on a part's routing or process sheet.

direct materials / Materials used in the manufacture of a part or product that are listed on the bill of materials.

direct view storage tube / A cathrode ray tube (CRT) that uses a grid behind the tube face to maintain an electronic picture. It does not use a refresh cycle, and it cannot show dynamics because whenever a part of the picture changes, the entire picture must be repainted.

discounted cash flow / A type of cash flow analysis where all future expenditures are converted by means of an interest factor or discount rate to their value in the time period the analysis is being made.

disk storage / Storing of binary data on rotating thin flat circular plates called disks using data tracks. The data are written on and read off the disks by movable read/write heads. Examples of disks used to store data are floppy, Winchester, or regular 14 inch disk drives.

dispatching / The short-term operational sequencing of jobs at individual work centers, including in large work centers the assignment of jobs to individual machines and work center personnel.

distributed processing / A form of data processing (DP) where the company's DP resources are scattered around several locations and the individual computer resources are linked by a communications network. In distributed processing, one computer can serve as the host or controlling computer for all the other computers in the distributed processing network.

DNC (direct numerical control) / NC machines controlled by a dedicated computer that stores many NC machine programs, schedules NC machine operation, downloads part programs to each machine tool, and tracks machine status fed back by each NC machine tool.

downtime / Any time when a machine or computer is not available for productive work owing to some sort of operating failure.

drop shipment / A method of shipment where shipments bypass intermediate suppliers and are shipped directly to a customer. The paperwork usually is sent through the intermediate supplier in such cases.

due date / The date an item is expected to be available for use.

emulate / To imitate the characteristics of one system with another system so that the results are transparent to the user.

end item / The finished good or product that is sold as an entity to a customer.

engineering change / A revision to the engineering specification of a product that must be reflected by a revision to the part's bill of materials, drawings, or specifications.

engineering change control or coordinator / The mechanism that determines the need and timing for an engineering change, and regulates the engineering change procedure by following policies established by the corporate engineering change committee.

expediting / The speeding up of a process so that it occurs in less than normal lead time; also commonly referred to as the deletion of red tape and the formal process to allow a task to be done quicker.

explosion / Refers to the multiplying of the lot size or required quantity times the bill of materials quantity to determine part requirements; commonly carried out at each level of the product structure or bill of materials.

fabrication / Refers to the building of subassemblies or components, as opposed to the assembly of the final product.

feedback / Data that come from the controlled process back to the control device or system so that a planned to actual comparison can occur, and corrective action can be taken if the actual differs from the planned or desired.

field / A part of a data record set aside for a specified element of data. The field is specified in terms of location and length within the record.

field or branch warehouse / A distribution point between the plant and the customer.

FIFO (first in–first out) / A method of inventory accounting where the first item in (or oldest shelf item) is used up first, and the inventory is costed for balance sheet purposes at the price of the latest items in stock.

file / A collection of data in the form of records that exists in a computer's memory or on some sort of storage device (disk or tape).

finished goods inventories / Inventory of end items available for customer sales. It can include spare parts.

finite capacity loading / Loading a factory or work center only to its capacity. This process automatically schedules lower priority items into the next available time period if the present time period's capacity is fully utilized.

firm planned order / A planned order where the start date, due date, or quantity is fixed by a planner, and cannot be altered by subsequent mrp runs.

firmware / Refers to instructions contained in read-only memory devices that cannot be changed by software command. Unlike "hardwiring," firmware can be changed by substituting another modular read-only memory device.

fixed order quantity / An ordering method where no matter what time period is involved, the order quantity remains the same; the order frequency then depends on demand for the part.

fixed period order / An ordering method where the *duration* of the time period for ordering is fixed, and the quantity needed or ordered over this time period can vary.

float / In CPM analysis, slack time in a parallel activity compared to an activity on a critical path.

flow shop / A mass production or continuous process environment where work usually flows through the same standardized layout of machines and processes—as opposed to job shop.

forecast / A prediction of future events using an algorithm that references historical data.

forecast error / The difference between the predicted or forecast value and the actual value for a time period.

FORTRAN (Formula Translator) / A high-level computer language usually used in scientific and mathematical work.

forward scheduling / Scheduling a process from a start date forward by operation time to arrive at a finish or due date.

full pegging / Refers to a process where pegging occurs all the way from the start level to the end item (parent) of the part, or the customer order being traced.

gateway work center / The work center where the first operation or actual direct labor first occurs on a part.

geometric modeling / Mathematically specifying a part or product by its geometric form or properties.

high-level language / A user-oriented computer programming language that must be converted to machine language by a compiler. Simplifies programming by creating a high-level command that equals a whole series of lower-level machine language instructions.

host computer / The computer that controls and schedules other computers in a distributed processing environment.

idle time / Lost production time on a machine due to setup, maintenance, or waiting for labor, parts, or tooling.

implosion / The aggregation of lower-level detailed data into a summary report; the reverse of explosion; commonly refers to the process of generating a where-used report.

indented bill of materials / A bill of materials that shows the composition of each subassembly by indenting the parts used in each higher-level part. Parts used more than once are not aggregated but are shown indented under their respective subassembly.

independent demand / Demand for a part or item not associated with any other part's demand.

indirect labor / Labor not directly associated with, or a part of, the production of any product—such as labor performed by a material handler, security guard, etc.

indirect materials / Miscellaneous materials used in the manufacturing of a product but not specified on the bill of materials or specifically figured in the part cost buildup. The cost of these materials is usually added to product cost as a part of overhead or miscellaneous manufacturing expense.

infinite capacity loading / Loading a plant or work center without regard to the capacity of that location. Used to show where overloads exist so that they can be corrected.

input / Work entering a work center. In data processing, data are input to storage devices (disk or tape or core memory) for later processing. When a computer program runs, data have to be input to the computer for manipulation by the CPU.

input/output control / Measuring a work center's (or device's) performance by comparing the rate at which work is coming into the work center against the rate the work is leaving the work center.

integrated circuit (IC) / Solid-state electronic circuits containing devices such as transistors, resistors, and capacitors that act together to form a complete circuit.

interactive / Refers to real-time use of a computer through a terminal where a programmer or person is "on-line" and conducts an active interchange via software with the computer.

inventory / A quantity of parts or products that serve as a buffer or intermediate supply that decouples two operations or a reserve stock of goods built up in anticipation of future demand. Since the inventory level can rise or fall, the operations on either side of the inventory no longer have a direct dependency on each other.

inventory turnover / How often in a period of time the inventory "turns over" or is used, calculated by dividing sales at cost by the average inventory dollar volume.

item / Refers to any discrete part, material, subassembly, or assembly.

job shop / A batch-oriented production shop capable of processing many different types of work, where each job typically flows through the shop in a different sequence and number of operations.

large scale intergration (LSI) / Integrated circuits (ICs) containing 100 or more gate equivalents or other circuitry of similar complexity.

lead time / The amount of time necessary to carry out a complete task. Can be comprised of paper work time, setup time, run time, queue time, move time, receiving and/or inspection time.

lead time offset / The amount of time between the date an item or activity is due and when it is started.

learning curve / A theory stating that in some production situations, each time that cumulative production volume doubles, the cost necessary to produce the items decreases by some percentage due to the workers' greater production experience.

level loading / Arranging the load on a work center or factory so that the loads per time period are approximately equal.

LIFO (last in–first out) / A method of inventory accounting where the last item in (or youngest shelf item) is used up first, and the inventory is costed for balance sheet purposes at the price of the oldest items in stock.

linear programming / A type of mathematical modeling where linear functions are optimized (either min or max) while bound by constraints. It usually involves thousands of calculations attempting to solve many simultaneous equations.

line balancing / Assigning tasks to workers and work stations on a production line so that work times and tasks are equalized, idle time is minimized, and the production can flow smoothly at a consistent rate.

load / The amount of work assigned into a work center or plant, expressed in labor hours or as a production rate.

load profile / The projected load pattern over several periods of time.

location code / The alphanumeric code that conveys the location of inventory in a stock room or warehouse.

lot size / The quantity of an item ordered from the shop or a vendor; may be more than the actual requirement quantity depending on lot-sizing rules.

low-level code / A code identifying the lowest level in a bill of materials where a given part number appears; holds that part out of the mrp netting process until all higher-level gross requirements for it have been summarized in the bill of materials explosion process.

machine center / A specific area or department of one or more machines that is considered as an entity for production planning and reporting purposes.

machine utilization / The amount of time (percentage) a machine is doing work compared to its total available time. Work time plus idle time equals available time.

macro / Short for macroscopic, such as a high-level or broad-brushed view.

make-to-order / Products manufactured after receipt of a customer's order specifying the final configuration and/or delivery time of the product.

make-to-stock / Products made to standard specification and placed in finished goods inventory.

manufacturing lead time / The complete time necessary to manufacture an item. Includes order paper work time, setup time, run time, queue time, move time, inspection time, stock pickout, and stocking time.

master production schedule / A statement of the products or parts the company is committed to manufacture. Lists specific part numbers, and quantities and due dates for each number.

material management / A concept of management that controls the complete flow of materials from purchase and production planning through finished goods delivery to the customer.

material requirements planning ("little mrp") / An order scheduling mechanism that sets and maintains order priorities for manufactured or purchased parts. The basis for the scheduling is the explosion of required product quantities (determined by the time-phased master production schedule) through each product's bill of materials to determine gross part requirements, and then netting these requirements against parts on hand and due in (scheduled receipts).

mean / The arithmetic average of a group of numbers.

mean absolute deviation (MAD) / The average of the absolute differences between the planned (or forecasted) and actual values in a group of numbers.

median / The central value of a group of numbers, i.e., the average of the two extreme values.

memory / Wherever data can be stored in a computer or peripherals. Can be core memory, disk, or tape. Memory can be read only, or changeable, i.e., read from/write into.

menu / A list of available responses or computer commands to be inputted as instructions in a computer program.

micro / Short for microscopic, such as a detailed or close-range view.

microcomputer / A small capacity computer, usually contained on one standalone circuit board, that contains one or more microprocessors, a memory of some type, and sometimes one or more peripheral devices.

microprocessor / A semiconductor device that performs arithmetic, logic, and decision-making operations as directed by a set of instructions.

min–max system / An ordering algorithm that orders the quantity needed to bring the inventory back to its maximum when the inventory quantity reaches a minimum.

mode / The value that occurs most often in a group of numbers.

moving average / A method that averages a group of the most recent *n* numbers by adding the latest observation and dropping the earliest observation.

modem / A device that changes digital data to analog data or vice versa so that the computer can transmit data to other computers over analog-based telecommunication lines.

NC (numerical control) / Controlling machines or machine tools by numerically coded commands that move machine parts along any of one or more axes.

nesting / Grouping patterns together in the most material-efficient manner for cutting; in computer programming, the placing of one "DO" loop within another.

net change mrp / Refers to transaction-driven materials requirements planning where the plan is updated by transactions such as inventory receivals/withdrawls, engineering changes, etc., that generate partial bill of materials explosions. Normally a completely new mrp plan is not (but can be) regenerated (as in the regenerative method).

net requirements / Part requirements after gross requirements are netted against on-hand inventory and parts due in (scheduled receipts).

network planning / Project planning and control through network procedures such as PERT and CPM, which seek to identify the controlling or limiting paths in a complex process.

normal distribution / A statistical distribution with a mean of zero and a standard deviation of 1, typified by a bell-shaped curve.

normal capacity / The maximum rate at which a task can be performed with standard manning and a standard work configuration.

object program / A computer program, usually in assembly or machine language, that is ready for execution by the computer. It is the result of running a source program through a compiler.

on hand / Present or available for use in inventory.

on-line / Refers to direct interactions with a computer where there is immediate response to an inputted command. Opposite of batch.

on order / Items currently on request from a supplier such as a vendor or manufacturer.

open loop system / A control system with no feedback loop.

open order / A released and unfilled order, either to manufacturing or purchasing.

operating system (OS) / Instructions in the form of programs that control the operation of a computer and its peripheral equipment (tape drives, disk drives, printers, terminals, etc.).

operations sequence / The step-by-step method by which a part is manufactured or a series of operations is carried out. Usually listed on a routing or process sheet.

order / A request for an item.

order backlog / The quantity of outstanding orders received but not shipped.

ordering cost / The cost of preparing an order. Includes clerical and paper work processing costs.

order point / The inventory balance below which an order for more parts is triggered.

order priority / The value or importance an order has in relation to other orders.

order quantity / The lot size or quantity ordered to replenish inventory. Usually determined by lot-sizing rules and subject to modification by a minimum, maximum, etc.

order release / The conversion of a planned order into a released (active) order that would go to manufacturing or purchasing. This action changes a planned order into a scheduled receipt.

overhead / General indirect costs that are aggregated and distributed to all units produced by some allocation method such as direct or absorption costing; also referred to as burden.

overlapping / A procedure where parts of a lot that are through operation 1 will be started on operation 2 before the entire lot is through operation 1.

part number / An alphanumeric code that specifically and exactly identifies a part as unique.

part program / A complete set of NC instructions to produce a part on an NC machine.

part programmer / An individual who programs NC machines.

past due / Late, delayed work or orders; work not done (or missed) in the period it was assigned to be done.

pegging / Refers to the process of showing the source of the original demand for a part or an order. Pegging differs from "where used" in that "where used" shows all possible sources of part demand, while pegging shows the specific source of the part order demand.

perpetual inventory / A record that tracks the on-hand balance of a part and all incoming and outgoing transactions for the part that affect its in-stock balance.

PERT (Program Evaluation and Review Technique) / Similar to CPM (critical path method) but introduces probabilistic concept to activity times.

physical inventory / Commonly refers to the process of taking an inventory or periodically counting all inventory items.

picking / Pulling parts from stock.

picking list / The paper work showing a list of all parts to be picked for an order.

piece parts / Individual or unit parts in inventory—not assemblies of subassemblies.

pixel / An element in a raster CRT picture that can be on or off and/or show color. Each pixel is like a tile in a mosaic. A 512 × 512 resolution screen contains 262,144 pixels.

planned order / The output of material requirements planning; a requirement for a certain quantity of a part by a certain date. Planned orders are subject to review by a planner before release and can change from one mrp run to another depending on changes in a part's inventory balance, MPS changes, scheduled receipt changes, etc.

planning horizon / The total time span over which planning is performed in an MRP system. The planning horizon in most cases should equal or exceed the cumulative lead time of any part, whether purchased or manufactured. It is usually at least 1 year long.

point-to-point control / NC where the tool is not continuously engaged with the workpiece. The work is moved to the correct position, the tool operates on the workpiece and is then withdrawn, and the workpiece is moved again.

postprocessor / The computer instructions that convert cutter location (CL) data, based on a part's geometry, to the proper specifications necessary to run a particular piece of NC machinery. Each type of NC machine has its own unique postprocessor.

priority rules / Rules which allow someone to know what job to do next, i.e., which job is most important in relation to other existing ones.

process sheet / The detailed part routing that shows each step in the manufacture of a product.

procurement / The act of obtaining items from outside vendors.

product mix / The various amounts of different products in the overall product line.

product structure / The linkage that establishes the relationship of all parts in a product's bill of materials.

product cycle / The total time needed to manufacture a product.

productivity / A measure of output relative to input for a process, i.e., a measure of efficiency.

program / A set of coded instructions for a computer to carry out certain arithmetic and logical operations.

purchase order / A document completely describing what a buyer is ordering from a vendor. It includes information on part descriptions, price, quantity, need date, shipping terms, credit terms, etc.

purchase requisition / An internal order to the purchasing department authorizing the purchase of a certain amount of any item by a certain date.

purchasing lead time / The complete time necessary to procure an item. Includes order paper work time, vendor lead time, shipping time, receiving and inspection time, and stocking time.

quality assurance (QA) / The function that has the responsibility to see that adequate product quality is ensured in the design process. Reflects the principle that quality is designed into a part or product.

quality control (QC) / The function that has the responsibility to see that parts and products are made exactly to the specifications shown on the engineering drawings.

queue / A line of goods or people waiting to be processed in some manner.

random / Occurrences that have no regular or forecastable pattern, where the chance of any one occurrence is no greater than that of any other.

random access memory (RAM) / A method of data storage where any element of data can be directly accessed equally as fast as any other data element.

raster scan display / A form of CRT display where a moving electron gun scans the TV screen in a regular line-by-line pattern lighting up programmed parts of the picture. In raster tubes, the picture is composed of a mosaic of tiles or squares called pixels. Resolution is determined by the number of pixels per square inch. Raster scan displays can be used for full color pictures and have excellent dynamic capabilities.

raw material / The basic stock (steel, aluminum, wood, etc.) purchased in order to make parts.

read-only memory (ROM) / A method of data storage where the data can be acessed to read only, and in normal operation cannot be changed or replaced by writing over it.

real-time operation / Computer processing or control performed at the same time as the controlled operations or process is occurring. Also accessing data that represent actual values at that point in time.

refresh display / A form of CRT display where only the line elements that make up a drawing are projected and connected on the face of the screen. This type of display must be "refreshed" 30 to 40 times per second to avoid flicker. It can show shades of color but not full color, and also has good dynamic capability.

regenerative mrp / A form of materials requirements planning (mrp) where the complete master production requirements are exploded periodically through each product's bill of materials and routings to form a completely new priority plan. All planned orders from the previous "regen" mrp run are cancelled and a fresh scheduling "snapshot" of labor and material requirements is produced.

release / To convert a planned order into an open order that is then known as a scheduled receipt.

remote job entry (RJE) / Controlling or executing computer operations from a terminal remote or far away from the computer itself through a communication line.

reorder point / See **order point.**

reorder quantity / See **order quantity.**

replenishment lead time / The total elapsed time from the decision to reorder a part until the part is in stock available for use.

requisition / An authorization for specific materials or parts to be obtained either from a parts vendor or the stock room in a specified quantity at a certain time.

resource requirement planning / "Rough cut" planning based on product family planning bills of labor, materials, or other resources performed to ensure that the production plan is achievable before it is disaggregated to form the detailed master production schedule. This resource planning does not take into account existing open shop orders.

response time / The elapsed time between the command to start any computer operation and the receipt of the desired operation's answer. Includes transmission time to and from the computer and the time of CPU processing.

routing / The list of instructions of sequential operations to manufacture a part. Usually showing each work station, operation, setup time, run time, and machines, tools, jigs, and fixtures necessary for each production step.

run time / The time required to make a complete execution of a single computer program.

safety stock / An extra supply of stock carried to protect against stockouts caused by fluctuations in demand.

safety time / An extra time allowance inserted ahead of the real due date to protect against fluctuations in supply.

scanner / A laser-operated device that can recognize bar code or alphanumeric font and can convert information from these items to digital data.

schedule / A list of tasks to be performed by a given work center or facility that connotes sequence and priority.

scheduling rules / Policies with which schedules can be created and maintained, such as how to give priorities to orders.

scrap allowance / A factor that is used to increase gross requirements for a part to compensate for scrap generated during the manufacturing process.

sequencing / Arranging the order in which tasks will occur.

service level / A measure of the effectiveness of customer service, usually expressed as a ratio or percentage such as percentage of line items shipped versus ordered.

setup cost / The cost of preparing a production machine to run many production parts.

setup time / The time it takes to prepare a production machine to run many production parts. Usually includes time necessary to get tools and fixtures, setup the first part, and perhaps even run and inspect the first part produced.

shop calendar / The establishment of an agreed upon company calendar denoting how the days of the week relate to production days when the factory is operating. Usually done on a daily basis starting at production day 1 of a year and ending on the last production day of the year, i.e., last day of 1979 = shop day 9250, first day of 1980 = shop day 0001.

shop order / An order calling for the manufacture of a quantity of items by a certain date according to its authorized process.

significant part numbers / Part numbers whose alphanumeric arrangement convey information about such things as product class, size, and material.

single-level pegging / A routine that traces requirements for a part or order up the product structure or order records one level at a time to reveal the source of any part or order requirement.

slack time / Any time remaining between the end date or estimated end date of a job and the due date.

smoothing / Averaging a series of numbers to remove wide fluctuations in the values. Can be by moving average technique or its more-sophisticated relative—exponential smoothing.

software / The set of coded instructions that contain logic and arithmetic directions for the manipulation of a set of data by a computer.

source language / The original high-level language in which software was written. It is converted to object code by a compiler. Examples of source language are BASIC, COBOL, and PASCAL.

split lot / A released shop order that has its original lot size quantity divided into two or more parts, usually in order to expedite work on one portion of the lot.

staging / Withdrawing and "kitting" (or assembling) of material to ensure its availability; informally used as a method to find out part shortages, list the shortages, then work the shortage list.

standards / The industrial engineering time allowances measured and permitted to be utilized as the basic rate for an operation.

standard costs / The cost of labor, material, or overhead that is established by accounting after historical analysis of past trends. Used for planning purposes and as a standard from which to track variances.

statistical inventory control / Inventory control using a statistical or mathematical analysis of past history to determine future inventory requirements, order points, etc.

stock out percentages / The number of line items stocked out (no on-hand balance) divided by the total number of order line items.

subcontracting / The farming out of manufacturing work to outside vendors. Usually utilized as a way of solving a capacity problem in a plant.

summarized bill of materials / A list of all parts that make up the parent product, where product structure information is not conveyed and parts that appear in several places are aggregated and listed only once.

time-phased planning / Planning future operations based on a series of related sequential time segments or buckets.

time sharing / The interwoven use of a number of computer terminals attached to one computer where any one user is under the illusion he or she is the only user; interactive use of a computer terminal.

tracking signal / The running sum of the forecast errors (RSFE) divided by the mean absolute deviation (MAD) of a series of numbers.

turning center / An NC machine capable of performing many lathe-oriented operations such a boring, facing, turning, and threading. Usually equipped with automatic tool changing, and a multitool holder.

turnkey system / A computer-operated system designed, built, installed, and tested by the vendor and turned over to the customer as a fully operational system.

turnover / A measure of stock usage arrived at by dividing sales at cost by the average inventory level in dollars at cost for a time period.

value added / A company's sales minus purchased parts and raw materials and outside services.

vendor lead time / The elapsed time between the placement of a purchase order to a vendor and when the goods are delivered and available for inspection and/or subsequent use.

weighted average / A method of averaging where the numbers to be averaged are factored or weighted by different values. The sum of weighting values must always equal 1.

where-used report / A report showing every part in the company's product line in which a certain part number is utilized, and the level on the part's bill of materials at which it is utilized.

work center / A place where people or machines are located to perform manufacturing operations.

work center efficiency / The ratio of standard or allowed hours divided by actual hours worked on an operation.

work in process / All materials, parts, and subassemblies that lie between release of the raw material to make a part and the finished product in finished goods inventory.

work order / An authorization document to allow labor to be performed on a part.

yield factor / The percentage of material output from a process to the material that went into the process.

Bibliography

American Production and Control Society, *Annual Conference Proceedings* (1974–1979). Washington, DC: 1974–1979.

American Production and Control Society, *Master Production Scheduling Reprints.* Washington, DC: 1977.

Brooks, Frederick P., *The Mythical Man-Month.* Reading, MA: Addison-Wesley Publishing Company, 1975.

Brown, Robert Goodell, *Materials Management Systems.* New York: John Wiley & Sons, 1977.

Corey, E. Raymond, *Procurement Management: Strategy, Organization, and Decision-making.* Boston, MA: CBI Publishing Company, Inc., 1978.

Enrick, Norbert L., *Market and Sales Forecasting,* 2nd rev. ed. Melbourne, Florida: Robert E. Krieger Publ. Co., 1980.

Harrington, Joseph Jr., *Computer Integrated Manufacturing.* Huntington, NY: Robert E. Krieger Publishing Company, 1979.

Lambert, Douglas M., *The Development of an Inventory Costing Methodology: A Study of the Costs Associated with Holding Inventory.* Chicago: National Council of Physical Distribution Management, 1975.

Lester, Ronald H., Enrick, Norbert L., and Mottley Jr., Harry E., *Quality Control for profit.* New York: Industrial Press Inc., 1977.

Makridakis, Spyros, and Wheelright, Steven C., *Forecasting—Methods and Applications.* New York: John Wiley & Sons, 1978.

Mather, Hal, and Plossl, George W., *The Master Production Schedule—Management's Handle of the Business.* 2nd ed. Atlanta: Mather & Plossl, Inc., 1977.

Orlicky, Joseph, *Material Requirements Planning.* New York: McGraw-Hill Book Company, 1975.

Plossl, George W., *Manufacturing Control—The Last Frontier for Profits.* Reston, VA: Reston Publishing Company, Inc., 1973.

Plossl, George W., and Welch, W. Evert, *The Role of Top Management in the Control of Inventory.* Reston, VA: Reston Publishing Company, 1979.

Smith, Bernard T., *Focus Forecasting—Computer Techniques for Inventory Control.* Boston: CBI Publishing Company, 1978.

Wight, Oliver, *Production and Inventory Management in the Computer Age.* Boston: Cahners Books International, Inc., 1974.

Index